FIDDLING WITH THE INFINITESIMAL

Fiddling With The Infinitesimal
A Guide Through Many Astonishing Results In
Foundational Calculus

By

Matthew Stephen Fox
Harvey Mudd College

Terra, 2016–2017

SAPERE AUDE

Copyright

Copyright © 2017 by Matthew Fox
All rights reserved.

First Edition: 2017

ISBN-13: 978-1973984399   ISBN-10: 1973984393

This book is protected under the copyright laws of the United States of America. No portion may be reproduced or used by any means whatsoever without the express written permission of the author with the exception of brief citation in a course setting or other scholarly environment.

Contact the author via infinitesimalfiddling@gmail.com

Figures produced by Mathematica and Ipe, document prepared by LaTeX

For Noni. My love for you reaches far beyond the remotest stars.

Preface

Mathematics, especially calculus, contains an abundance of unexpected and counterintuitive formulations. Why not write a book devoted to deriving and exploring some of these equations? Hence the present text.

FIDDLING WITH THE INFINITESIMAL is to be treated like a handbook, a manual of sorts, that guides the reader through the many engrossing applications and beautiful ideas of basic calculus. In this sense the book need not be read all at once, but rather referenced periodically for when one craves an enticing mathematical derivation.

As with any mathematics text, a few prerequisites are suggested before diving in. The architecture of this book resembles that of most introductory courses in calculus, wherein students encounter limits, derivatives, integrals, and series. While a thorough understanding of these topics is not necessary, elementary comprehension is assumed. An additional understanding of products is beyond sufficient, and hence not at all necessary. All readers are assumed proficient in algebra, geometry, and trigonometry.

Each chapter begins with a rudimentary review of the topic being pursued. For instance, the second chapter (on differential calculus) reviews the notion of a limit, conditions for existence, a few definitions, and the omnipotent derivative. Though informative, this book is no substitute for a calculus textbook. What we have written here is merely a casual textbook companion—an informal work filled with exuberant computations designed to make the most tranquil students of science and mathematics giddy.

The book begins with a brief note on mathematical beauty, notation, and an introduction to what we call the "claim-proof" writing structure of mathematics. This, like algebra to calculus, provides the reader with some foundational ideas and notation so the text is more readily understood. The second chapter deals with differential calculus in which the notions of limits and derivatives are ubiquitous. Here, counterintuitive constructions such as the Dirichlet function and Koch snowflake are explored. The third chapter concerns the fascinating features of integral calculus, including prominent functions such as the gamma and Gaussian functions as well as a proof on the irrationality of π. The fourth chapter extends into the field of mathematical analysis in which we explore many unexpected sums and products for π and Euler's constant e. Additional sections include a proof on the irrationality of e and the esteemed Riemann zeta function. The final chapter ventures into fields beyond calculus, such as number theory, combinatorics, and mathematical physics. Here, topics span from the Fibonacci sequence to black holes.

A student who has had, or who is taking, an introductory calculus course should be successful in handling the material throughout this book. That said, reading mathematics often requires an extensive amount of work to truly grasp the concepts. Hence, those with the above credentials must also possess a strong work ethic to honestly understand the ideas. In general, those who work hard are the ones who learn the most.

Acknowledgements

I am beholden and incredibly grateful to Matthew Calligaro and Ranganath Selagamsetty. Besides their talent for detecting mathematical and grammatical errata, their thorough revisions and constructive criticisms significantly crafted the present text. I thank Jacob Fox for his help reformulating the introduction, and Cache Sanchez for listening to my frequent, aloud revisions.

The reality of this book is due to Chuck Hendrick and Janie Mueller. My deep revere for their relentless benevolence and insatiable curiosity ultimately morphed into the inspiration for this work. Incalculable thanks to my professors, my peers, and, certainly, my family.

Table of Contents

Preface — vii

Acknowledgements — ix

1 Introduction — 4
   1.1 Some Notation 6
   1.2 The Claim-Proof Structure 10

2 Limits and Differential Calculus — 18
   2.1 Antipodal Temperatures and Wobbly Tables 23
   2.2 Euler's Number 29
   2.3 An Arduous Limit 32
   2.4 The Dirichlet Function 38
   2.5 The Koch Snowflake 44
   2.6 Kelly's Criterion 52

3 Integral Calculus — 56
   3.1 The Quadratic Formula 63
   3.2 Euler's Equation and the Proverbial $e^{i\pi}$ 64
   3.3 The Parabolic Nature of Free Fall 67
   3.4 Gabriel's Horn 75
   3.5 Coins and Cycloids 79
   3.6 The Gaussian Function 86
   3.7 The Gamma Function 95

3.8	The Catenary	102
3.9	Buffon's Needle Problem	109
3.10	On the Irrationality of π	114

4 Series and Products — **122**

4.1	Taylor Series and π	127
4.2	Euler's Equation Revisited	131
4.3	The Harmonic Series	134
4.4	The Birthday Paradox	140
4.5	On the Infinitude of Primes	145
4.6	On the Irrationality of e	150
4.7	The Basel Problem	153
4.8	The Wallis Product	159
4.9	Riemann's Rearrangement Theorem	165
4.10	Viète's Formula for π	171
4.11	The Riemann Zeta Function	175

5 Beyond Calculus — **186**

5.1	Algebra: Nested Radicals and Tetration	186
5.2	Number Theory: The Golden Ratio	191
5.3	Combinatorics: Pascal's Triangle	197
5.4	Set Theory: $\infty > \infty$	205
5.5	Multivariate Calculus: A Hole in the Earth	212
5.6	Mathematical Physics: Relativity	216
5.7	Research Mathematics: Open Conjectures	226

Appendix — **234**

A.1	On the Irrationality of $\sqrt{2}$	234
A.2	A Note on Geometric Series	235
A.3	Derivation of Newton's Law of Gravity	236
A.4	Digression on the Maclaurin Series for e^x	242
A.5	Reconciling Two Formulas for e	243

A.6 The Fundamental Theorem of Arithmetic 245
A.7 On Grandi's Series 247
A.8 Proofs by Induction 250
A.9 The Differential Equation $ay'' + by' + cy = 0$ 253
A.10 Mersenne Primes and Perfect Numbers 257

Algorithms for Select Conjectures   260

Selected Bibliography   263

List of Symbols   266

Index   268

Chapter **1**

Introduction

"The mathematical sciences particularly exhibit order, symmetry, and limitation; and these are the greatest forms of the beautiful."
  ∼ Aristotle

Similar to the finest telescopes, mathematics provides a detailed look into the intricate structure of the universe. The curiosity of the human spirit inevitably forces us to consider the unanswerable: Why should mathematics provide such a lens? Why should mathematics be so successful at describing impossibly ornate constructions like black holes and subatomic particles well before humanity is even capable of observing such objects? In this way, mathematics resembles a deistic clairvoyant able to foresee profound cosmogonic theories, each superseding its predecessor, and each progressing towards the true explanation for our existence.

The inexorable consistency, the simple lack of contradiction, has coaxed many to interpret mathematics as an encyclopedic description of the universe—*the* language in which the laws of the greater cosmos are written. And some go further, philosophizing mathematics as the demiurge to whom we owe our being. Though some are rightfully atheistic to such a sermon, all mathematicians will agree on the resplendent beauty embedded within the arcane notation and abstract theorems. It has been said that the experience of a great proof in mathematics is comparable to a profound physical beauty, akin to a dopamine surge from sensual interaction. While this equivalence is not readily verifiable, the analogy corroborates the point: Mathematics, like art and literature, is a winsome establishment capable of entertaining the human mind in the most pleasurable of ways.

Now many non-mathematicians may disagree. To them, the study seems a monotonous, esoteric portion of academia pursued by only the most numerically inclined scholars. While this holds historic credence, contemporary mathematics is unknowingly accessible to all who house proper discipline and who are willing to approach the symbols without fear. Like many academic pursuits, mathematical ones guarantee a certain level of satisfaction. This can be obtained from, say, a simple yet interesting fact (e.g. there are about 10^{120} different games of chess but only 10^{80} atoms in the observable universe) or a detailed comprehension of the equations governing something familiar (e.g. the dynamics of an airplane in terms of Newton's $F = ma$).

Indeed, functions and computations are the very fabric on which many of our daily activities rest. Consider, for example, your mobile phone. This device is nothing but a minute (albeit powerful) computer whose history (and perhaps existence) would be vastly altered had mathematician Alan Turing pursued a degree in business instead. Moreover, we take it for granted that our phones are aware of their position on Earth, enabling a map to be drawn from your current position to, say, the nearest Starbucks. This mapping feature, a consequence of the Global Positioning System (GPS), we owe to the formulae of Albert Einstein and Isaac Newton.

Yet this versatility stems farther than just your phone. Onto just a napkin, the studious mathematician can jot down the details of Earth's orbit about the Sun or, just as easily, illustrate how casinos withdraw so much money from determined gamblers. This aspect of mathematics, its universal applicability, is the first of two principal beauties we acknowledge in this book. The second is mathematics' internal architecture, its ability to bring two seemingly unconnected entities into equality, such as π and four times the boundless sum

$$1 - \frac{1}{3} + \frac{1}{5} - \frac{1}{7} + \frac{1}{9} - \frac{1}{11} + \frac{1}{13} - \frac{1}{15} + \cdots.$$

We intend to delineate many of the most famous relationships, the above sum among them, derivable through techniques learned in any introductory calculus course. We hope that when the final page is turned, your aura is gleaming with a novel appreciation of the many surprises in mathematics, especially those in calculus. And if not this, at least you'll be aware of a few equations guaranteed to impress your friends.

§1.1 Some Notation

As one progresses to more advanced mathematics courses, mathematical notation appears to supersede common language. Though not entirely true, notation does become more complex and the corresponding ideas more abstract. Such is substantiated by more formal mathematical texts, which are often riddled with symbols one may struggle to transcribe, and ideas one may find hard to envision (for instance, what does a seven-dimensional sphere look like?). There are some symbols, however, that are universal throughout the field (such as $=$), rendering them more substantial than others. These symbols make the communication of mathematical concepts more effective and succinct. It is the purpose of this section to specify these symbols so we may exploit them later on.

§1.1. Some Notation

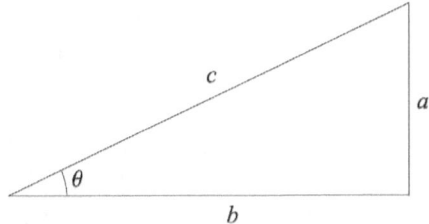

Figure 1.1: A right triangle with hypotenuse c and legs a and b.

As you know, the equals sign $=$ indicates when two mathematical expressions are identical.† It is often useful, however, to go a bit further and say that, *by definition*, expression A is expression B. Though this can be done in a myriad of ways, we will utilize the symbol $:=$.‡ So when we write

$$A := B,$$

we are saying that A equals B and also that A is defined by whatever B is. Though abstract, you are likely familiar with a few examples of this notation. Consider the famous number π—the constant *defined* by the ratio of a circle's circumference C to its diameter d. Symbolically, we write

$$\pi := \frac{C}{d}$$

from which the reader can conclude that π not only equals $\frac{C}{d}$, but is defined by the ratio $\frac{C}{d}$.

Perhaps a more sophisticated example would be the trigonometric functions: sine, cosine, and tangent. Using the triangle in Figure 1.1,

$$\sin(\theta) := \frac{a}{c}, \cos(\theta) := \frac{b}{c}, \text{ and } \tan(\theta) := \frac{a}{b}.$$

---

†If this is new to you, it's probably best to stop reading now.

‡Other common alternatives include the symbols: $\equiv$, $\stackrel{\text{def}}{=}$, and $\stackrel{\Delta}{=}$ (the last read "delta-equal").

Similar to the π example, these expressions are more than equalities, they are the definitions of the trigonometric functions, hence the use of $:=$ over the familiar $=$.

Another ubiquitous symbol is $\implies$, with is mathematics for "implies." Thus, writing $A \implies B$ conveys that A implies B, and should be read as such. To illustrate, suppose we have the equation $A + B = C$ and want to relay the implication that $A = C - B$. Rather than using words, we can achieve our goal by writing

$$A + B = C \implies A = C - B. \tag{1.1}$$

Indeed, this is a cleaner and more succinct means for detailing the implication than by using words.

In the previous paragraph, take note of the (1.1) following the centered equation. This parenthetical allows us to reference the equation juxtaposed to it. In this book, such referencing is used extensively so that previous equations can be mentioned without hassle.

Another popular character is $\in$, which crops up frequently when talking about sets. Recall that a *set* is a collection of distinct objects, such as the set

$$S := \{1, 2, 3, 9\},$$

which is a collection of the numbers $1, 2, 3$, and 9. Each object in a set is called an *element*, so the numbers $1, 2, 3$, and 9 are the elements of S. The symbols $\in$ and $\notin$ are used for describing elements in and not in particular sets, respectively. To illustrate with S above, we would write $1 \in S$ to express that one is in S and write $42 \notin S$ to indicate that forty-two is not in S. This notation is particularly useful for detailing the domain of functions. For example, consider the function $g(\theta) := \cos^2(\theta)$ whose domain consists of the elements in S. Rather than explaining this restricted domain in words (as we

§1.1. Some Notation

have just done) we can convey the same information by writing

$$g(\theta) := \cos^2(\theta), \text{ where } \theta \in \mathcal{S}.$$

This notation is also used in expressing a function's domain via *interval notation*. Now an interval is indeed a set, except it is a continuum, like a portion of the number line, and not composed of discrete values as in $\mathcal{S}$ above. So if $h(x)$ is a function defined for all x such that $1 \leq x \leq 2$, we write $x \in [1, 2]$ to convey that x is some real number between one and two, inclusive (this is the interval for x on which $h(x)$ is defined). To exclude endpoints (e.g. the values one and two) we use parentheses rather than brackets. So if instead $1 < x < 2$, we write $x \in (1, 2)$. Naturally, it may be that $1 \leq x < 2$ or that $1 < x \leq 2$. These are no problem, we simply use both parentheses and brackets in the notation: $x \in [1, 2)$ or $x \in (1, 2]$, respectively. The general rule of interval notation is to always associate parentheses with $<$ and $>$ and brackets with $\leq$ and $\geq$.[†]

We now transition the discussion towards a few sets which are ubiquitous throughout mathematics—the first being the set of *natural numbers*, denoted $\mathbb{N}$. The natural numbers consist of all the counting numbers: $1, 2, 3, 4, \cdots$. Note that zero is not considered a natural number—that is, $0 \notin \mathbb{N}$. The set of natural numbers union zero is called the set of *whole numbers*, denoted $\mathbb{N}_0$. This is the set consisting of the numbers $0, 1, 2, 3, \cdots$. The set of whole numbers union the additive inverse of each element—that is, the number $-a$ such that $a + (-a) = 0$—is called the set of *integers* and is denoted $\mathbb{Z}$. This is the set composed of the numbers: $\cdots, -2, -1, 0, 1, 2, \cdots$. The set of *prime numbers*, denoted $\mathbb{P}$, is a subset of $\mathbb{N}$ and includes

---

[†]Note that $\pm\infty$ are not bona fide numbers—infinity is merely the notion of never ending. Hence, it makes no sense to say $-\infty \leq x \leq \infty$ because x can never equal $\pm\infty$. Therefore, we always write $-\infty < x < \infty$, which, in interval notation, uses parentheses: $x \in (-\infty, \infty)$.

only those elements whose divisors are one and itself. For instance, five is prime because five is evenly divisible by only one and five. Conversely, four is not prime, making it *composite*, because four is divisible by one, two, and four. We show in the appendix that primes are the building blocks of all integers—they are, in some sense, the atoms of the number line. Other prime numbers include: 2, 3, 7, and 11.† The next set concerns *rational numbers*, which are numbers that can be expressed as a fraction $\frac{a}{b}$, where a and b are integers—that is, $a, b \in \mathbb{Z}$—and $b \neq 0$. The set of rational numbers is denoted $\mathbb{Q}$, contains all integers, and also contains fractions such as: $\frac{1}{2}$, $-\frac{101011}{42}$, and $\frac{42}{43}$. The set of rational numbers union all numbers which are not rational, i.e. *irrational numbers* (such as π), is called the set of *real numbers* and is denoted $\mathbb{R}$. Some elements from this set include: $1, \frac{2}{3}, \sqrt{2}$, and -2.7182818. Finally, the set of *complex numbers*, denoted $\mathbb{C}$, is the set of real numbers union all numbers which are not real. A number which is not real is sometimes called an *imaginary number*. The unit imaginary number (analogous to the unit real number 1), denoted i, is defined by

$$i := \sqrt{-1}.$$

Complex numbers are all numbers of the form $a + bi$, where a and b are real numbers—that is, $a, b \in \mathbb{R}$. Some examples of complex numbers include: $-12, 2 + \frac{2}{3}i$, and $-\pi i$.

For reference, all the sets we've discussed are tabulated in Table 1.1.

§1.2 The Claim-Proof Structure

More formal texts in mathematics tend to follow a claim-proof writing structure intended to display an author's theory and proof

---

†For reasons justified later in §A.6, one is not typically considered prime—that is, $1 \notin \mathbb{P}$.

§1.2. THE CLAIM-PROOF STRUCTURE

$\mathbb{C}$	the set of complex numbers (e.g. $\pi, i, -2 + \frac{3}{4}i$)
$\mathbb{R}$	the set of real numbers (e.g. $1, \pi, -\frac{42}{3}$)
$\mathbb{Q}$	the set of rational numbers (e.g. $1, \frac{1}{9}, -\frac{12}{11}$)
$\mathbb{Z}$	the set of integers (e.g. $0, -2, 5$)
$\mathbb{N}_0$	the set of whole numbers (e.g. $0, 1, 2$)
$\mathbb{N}$	the set of natural numbers (e.g. $1, 2, 3$)
$\mathbb{P}$	the set of prime numbers (e.g. $2, 3, 11$)

Table 1.1: Common numerical sets.

as clear and concise as possible. Of all possible claims, mathematicians have categorized them into five different assertions: lemmas, propositions, theorems, corollaries, and conjectures. We will explain each shortly with a definition and example, but before we do there is one more symbol to go over, related to the acronym QED, which abbreviates the Latin phrase *quod erat demonstrandum* ("which was to be demonstrated").

QED is frequently found at the end of mathematical proofs so the reader knows when the proof is complete. That said, more modern mathematical papers prefer the box symbol □ over QED.[†] Whenever we decide to use the claim-proof structure, we will employ this contemporary notation; therefore, whenever you see □, it means the proof directly preceding it is finished.

Our conversation on mathematical arguments begins with lemmas, propositions, and theorems as these are closely related. *Lemmas* are assertions that an author proves so more substantial claims can be upheld. Think of lemmas as steppingstones to larger and more profound results, which are theorems and propositions. *Theorems* are regarded as the chief results in a paper—each a Holy Grail, so to speak. Though significant, *propositions* are seldom considered

---

[†]The □ is sometimes called a *tombstone*.

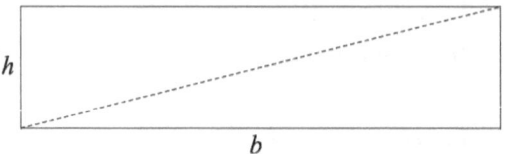

Figure 1.2: A rectangle with height h and base length b.

the main results of a paper—they lie in between lemmas and theorems on the importance hierarchy. That said, distinguishing between propositions and theorems can be a subjective task, so the exact labeling is left to the author.

Corollaries and conjectures are less ambiguous then propositions and theorems. A *corollary* is an assertion whose proof follows almost immediately from a theorem. Think of it as an interesting extension to a theorem whose proof is nearly identical to that of the theorem it follows. A *conjecture*, on the other hand, is a sort of hypothesis—a mathematical question which the author cannot rigorously solve but can (at least mildly) intuit.

To illustrate these more concretely, journey with us as we prove the Pythagorean theorem, among many other things. We start with the following lemma:

Lemma 1.1. *The area of a right triangle with height h and base b is $\frac{1}{2}bh$.*

Proof. To prove this lemma, consider Figure 1.2. As we see, the dashed line along the diagonal of the rectangle constructs two identical right triangles. Because the area of any rectangle with height h and base b is simply bh, it follows that the area of each right triangle is half this value.[†] □

[†]Two notes on reading proofs: (1) Do not despair if you must reread a proof to understand its logical structure. This is a common occurrence, especially when proofs become more intricate and concise. (2) After reading a proof, it is remarkably

§1.2. THE CLAIM-PROOF STRUCTURE

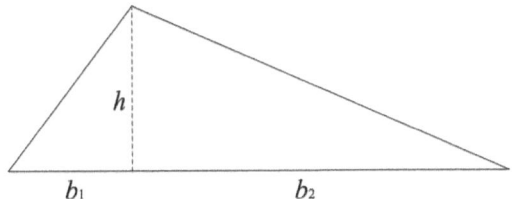

Figure 1.3: A triangle with height h and base length $b_1 + b_2$.

We have shown through a geometric argument that the area of any right triangle is indeed $\frac{1}{2}bh$. Notice that the proof of our initial claim (the area of any right triangle is $\frac{1}{2}bh$) is complete once the □ is met.

As you know, the right triangle is a special case in the set of all possible triangles. Might the area formula $\frac{1}{2}bh$ be true for *all* triangles? Though this is not our goal (we want the Pythagorean theorem) we can include it as a proposition:

Proposition 1.1. *The area of any triangle with height h and base b is $\frac{1}{2}bh$.*

Proof. To establish the proof, refer to Figure 1.3. Observe that the dashed line intersects the base at a right angle, thus creating two right triangles with base lengths b_1 and b_2. From Lemma 1.1, we know the area of each right triangle is $\frac{1}{2}b_1h$ and $\frac{1}{2}b_2h$, respectively. The area of the entire triangle is then

$$\frac{1}{2}b_1h + \frac{1}{2}b_2h = \frac{1}{2}(b_1 + b_2)h,$$

---

worthwhile to review the large scale structure of the proof and ask why the approach the author took substantiates the corresponding claim. While this is not necessary for these first few proofs (they are relatively trivial), it is vital to later, more complex ones. In this book, there will be no shortage of *how* to prove claims, but there may be a lingering itch for *why* particular proofs work. This end-of-proof reflection forces you to understand the *why*.

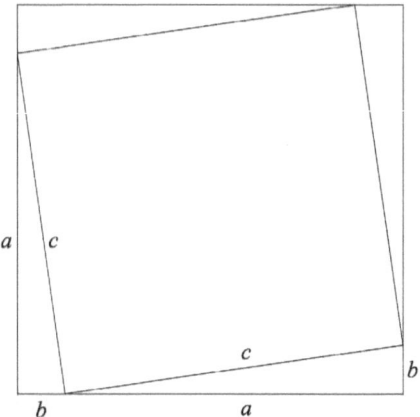

Figure 1.4: A square of side length c inscribed inside a square of side length $a+b$. (*Note*: The apparent tilt of the outer square is an optical allusion!)

which is the desired result. □

We have shown that *all* triangles have area $\frac{1}{2}bh$, where b is the base length and h the height of the triangle. As one can imagine, this result carries much more value than the previous lemma, which is why it's denoted as a proposition. To emphasize, this is not introduced as a theorem because it is a side-result in our pursuit of proving the Pythagorean theorem.

Theorem 1.1 (The Pythagorean Theorem). *For any right triangle with catheti a and b and hypotenuse c, $a^2 + b^2 = c^2$.*[†]

Proof. Consider Figure 1.4. Notice that the area of the outer square is $(a+b)^2$. From Lemma 1.1, the area for each of the four right triangles is $\frac{1}{2}ab$, making the area of the outer square also equal to

---

[†]The term *catheti* (singular *cathetus*) refers to the non-hypotenuse legs on a right triangle.

§1.2. The Claim-Proof Structure

$c^2 + 4(\frac{1}{2}ab)$. Thus,

$$(a+b)^2 = c^2 + 4\left(\frac{1}{2}ab\right)$$
$$a^2 + 2ab + b^2 = c^2 + 2ab$$
$$a^2 + b^2 = c^2,$$

which is the Pythagorean theorem. $\square$

We have successfully proven the Pythagorean theorem. Notice that in order to do so we had to use a less-significant result (namely Lemma 1.1) to arrive at the main punch line. Along the way, we also proved that all triangles have area $\frac{1}{2}bh$, which is a nice touch. A corollary to the Pythagorean theorem might be:

Corollary 1.1. *There exist infinitely many integer solutions in a, b, and c to the equation $a^2 + b^2 = c^2$ and they are given by*

$$\underbrace{(m^2 - n^2)^2}_{a^2} + \underbrace{(2mn)^2}_{b^2} = \underbrace{(m^2 + n^2)^2}_{c^2},$$

where m and n are arbitrary positive integers.

Proof. The proof is a simple algebraic expansion to verify the above equality:

$$\begin{aligned}(m^2 + n^2)^2 &= (m^2)^2 + 2m^2n^2 + (n^2)^2 \\ &= (m^2)^2 + 2m^2n^2 + (n^2)^2 + 2m^2n^2 - 2m^2n^2 \\ &= [(m^2)^2 - 2m^2n^2 + (n^2)^2] + 4m^2n^2 \\ &= (m^2 - n^2)^2 + (2mn)^2,\end{aligned}$$

which is the proposed formula. The infinitude of solutions follows from the arbitrary nature of m and n. $\square$

This is a powerful corollary to the Pythagorean theorem. Not only is $a^2 + b^2 = c^2$ true for all right triangles, but infinitely many of these triangles have integer lengths for all their sides given by

$$a = m^2 - n^2, b = 2mn, \text{ and } c = m^2 + n^2,$$

where $m, n \in \mathbb{N}$.

As interesting as this appears, the curious mathematician may still inquire about integer solutions to the related equation $a^3 + b^3 = c^3$. But after much calculation, no three integers a, b, and c seem to satisfy the expression. Out of frustration, the mathematician might move onto the fourth-order expression $a^4 + b^4 = c^4$ and again search for integer solutions in a, b and c. But the search fails again, and the mathematician struggles to find even one combination of a, b, and c that satisfies $a^4 + b^4 = c^4$. At a loss, the mathematician settles for the following conjecture—a question or hypothesis to which the mathematician does not have a sufficient answer, but suspects to be true:

Conjecture 1.1 (Fermat's Conjecture). *The expression $a^n + b^n = c^n$ has no integer solutions in a, b, and c for $n > 2$.*

This conjecture, called *Fermat's Conjecture* (frequently called *Fermat's Last Theorem*, after the seventeenth century mathematician Pierre de Fermat), is one of the most famous problems in the history of mathematics. We have presented it here as a conjecture, implying the solution is not yet known, but this is no longer the case. In 1994, mathematician Andrew Wiles announced his now famous (though quite involved) proof of the conjecture.[†] Due to its complexity, Wiles' proof is well-beyond the scope of this book, so we do not include it here. You'll just have to trust us that Conjecture 1.1 is, in fact, true.

Out of respect for the above conjecture, we include a seemingly unrelated corollary:

[†]Technically speaking, Conjecture 1.1 should be regarded as a theorem because it has been shown to be true. But this would ruin the flow of things.

Corollary 1.2. *If n is a positive integer greater than two, then the nth root of two is irrational. That is, $\sqrt[n]{2} \notin \mathbb{Q}$ for $n > 2$.*

Proof. By way of contradiction, suppose $\sqrt[n]{2} \in \mathbb{Q}$. Then $\sqrt[n]{2}$ is rational and can, by definition, be expressed in the form

$$\sqrt[n]{2} = \frac{a}{b}, \tag{1.2}$$

where a and b are integers with $b \neq 0$. Raising each side to the nth power, we have

$$2 = \frac{a^n}{b^n} \implies b^n + b^n = a^n.$$

Because $n > 2$, by Fermat's Last Theorem no integer solutions exist to this equation. This is a contradiction because we assumed $\sqrt[n]{2}$ could be expressed as in (1.2) for $a, b \in \mathbb{Z}$. Consequently, $\sqrt[n]{2}$ must be irrational for $n > 2$.[†] □

We have seen the progression of a simple geometric argument to a multitude of astounding conclusions. From deriving a formula for the area of a right triangle, we were able to prove that all triangles have area $\frac{1}{2}bh$, prove the Pythagorean theorem, find infinitely many integer solutions to the equation $a^2 + b^2 = c^2$, establish one of the most famous problems in all mathematics, and prove that the nth root of two is irrational for $n > 2$. This is certainly a variety, and is a neat demonstration of the versatile nature of mathematics. On top of this, we have acquired a new set of mathematical notation intended to facilitate our understanding of the material throughout this book.

Let's now journey into the realm of one of mathematics' most alluring fields: *Calculus*—the Latin word for the small pebble found on an abacus (and also the subject of this book).

[†] Quite rightly, you may be inquiring about the case where $n = 2$. Unfortunately, Fermat will not warrant a proof for $n = 2$ because we require $n > 2$ (why?). Refer to §A.1 in the appendix for the specific case when $n = 2$.

CHAPTER 2

LIMITS AND DIFFERENTIAL CALCULUS

"Nothing takes place in the world whose meaning is not that of some maximum or minimum."

∼ Leonhard Euler

LIMITS are of fundamental importance to all things calculus. Without them, the very foundation on which calculus is built would crumble like a sandcastle in water. By definition, a *limit* is the value to which a function approaches as its input tends toward a particular value. For instance, the function $f(x) := \frac{1}{x^2}$ advances toward $\frac{1}{4}$ as x approaches 2, and $\sin(x)$ goes to 1 as x approaches $\frac{\pi}{2}$. We can write these a bit more succinctly using limit notation:

$$\lim_{x \to 2} f(x) = \frac{1}{4} \text{ and } \lim_{x \to \frac{\pi}{2}} \sin(x) = 1,$$

where each $\to$ is read "approaches."

CHAPTER 2. LIMITS AND DIFFERENTIAL CALCULUS

Notice that for $f(x) := \frac{1}{x^2}$,

$$\lim_{x \to 2} f(x) = f(2) = \frac{1}{4}.$$

In other words, the limit of $f(x)$ as $x \to 2$ equals $f(x)$ evaluated at $x = 2$. Functions like this—whose limit at some point a equals the function evaluated at a—are said to be *continuous* at a. In more mathematical terms, a function is continuous at the point $x = a$ if as $x \to a$, $f(x) \to f(a)$.

This definition, like many in mathematics, can generalize quite easily. In particular, it generalizes beyond points and to the functions themselves. So for a function $f(x)$ that is continuous at all points (excluding the endpoints) in the closed interval $I = [a, b]$, we say $f(x)$ is a *continuous function* on I. In other words, $f(x)$ is a continuous function on $[a, b]$ if for every $c \in (a, b)$

$$\lim_{x \to c} f(x) = f(c). \tag{2.1}$$

Of course, such an equality need not be true. The class of functions for which (2.1) is not true on some open interval (a, b) are called *discontinuous functions* on $[a, b]$. Such functions can be hard to come by. Typically when we say a function is discontinuous, we mean it is discontinuous at a point in some interval—a point discontinuity—and not discontinuous at all points in the interval. In part, point discontinuities are what make limits such a useful notion.

Consider, for example, the same $f(x) := \frac{1}{x^2}$ as above, but this time in the limit as $x \to 0$. Unlike when $x \to 2$, it is the case here that

$$\lim_{x \to 0} f(x) \neq f(0).$$

This is because $f(0) = \frac{1}{0^2} = \frac{1}{0}$ is undefined.[†] Hence we say $f(x)$ is discontinuous at the value $x = 0$. And so for any interval $I = (a, b)$

[†]You may be inclined to say $\frac{1}{0} = \infty$. But then we have $\frac{2}{0} = \infty$, and so $\frac{2}{0} = \frac{1}{0} \implies 2 = 1$, which is clearly absurd.

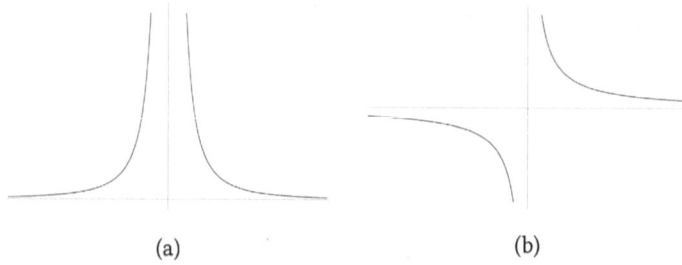

Figure 2.1: Graphs of (a) $\frac{1}{x^2}$ and (b) $\frac{1}{x}$ near $x = 0$.

for which $0 \in I$, $f(x)$, being continuous everywhere else on I, is discontinuous at the point $x = 0$—a point discontinuity.

That said, we can still deduce to what value $f(x)$ is approaching as $x \to 0$—it's infinity. Thus, even though $f(0)$ is undefined, the limit of $f(x)$ as $x \to 0$ is well-defined at ∞. That is,

$$\lim_{x \to 0} f(x) = \infty. \tag{2.2}$$

This shows that limits allow us to assign a value to a function at a point where the function is otherwise undefined. Make no mistake, however, the equality in (2.2) does not imply $f(0) = \infty$. And in general, for any function $g(x)$, the equality

$$\lim_{x \to c} g(x) = L \text{ where } L \in \mathbb{R}$$

does not imply $g(c) = L$. All the limit tells us it what a function looks like in the neighborhood of the value c as $x \to c$.

We should note that, like functions, limits are not always defined. For example,

$$\lim_{x \to 0} \frac{1}{x}$$

is undefined. Now this might appear odd, because we showed the limit of the companion function $\frac{1}{x^2}$ as $x \to 0$ does exist. This difference in existence has to do with how one approaches $x = 0$ between the two functions. Looking at the graph of $\frac{1}{x^2}$ in Figure 2.1(a), notice

CHAPTER 2. LIMITS AND DIFFERENTIAL CALCULUS

how the function approaches $+\infty$ irrespective of how one advances toward the origin. Conversely, for $\frac{1}{x}$ in Figure 2.1(b), approach the origin from $x > 0$ and the function heads off towards $+\infty$, while an approach from $x < 0$ causes the function to head off towards $-\infty$. This implies that the function approaches two different values at the same point, so there is no clear way to assign one value to the limit. We therefore say the limit of $\frac{1}{x}$ as $x \to 0$ does not exist.

This analysis permits us to define an existence condition for a limit. If c^+ denotes approaching the value $x = c$ along the x-axis from $x > c$ and c^- from $x < c$, then the limit of a function $f(x)$ as $x \to c$ exists provided

$$\lim_{x \to c^+} f(x) = \lim_{x \to c^-} f(x) = L.$$

Under this condition, we say

$$\lim_{x \to c} f(x) = L$$

and so the limit exists. If

$$\lim_{x \to c^+} f(x) \neq \lim_{x \to c^-} f(x),$$

then the limit does not exist.

We will now apply this idea to a particular limit that is ubiquitous throughout calculus—namely

$$\lim_{\Delta x \to 0} \frac{f(x + \Delta x) - f(x)}{\Delta x}, \tag{2.3}$$

where $f(x)$ is any continuous function. We say the function $f(x)$ is *differentiable* on $[a, b]$ if the limit in (2.3) exists for all values $x \in (a, b)$. This definition brings us to the broader definition of a *derivative*—something you've probably heard has something to do with slopes of functions. Before diving into any geometric meanings, however, it's worth mentioning that (2.3) is the definition of the derivative. Hence, the derivative of $f(x)$, denoted $\frac{d}{dx}f(x)$ or, equivalently, $f'(x)$, is

$$\frac{d}{dx}f(x) = f'(x) := \lim_{\Delta x \to 0} \frac{f(x + \Delta x) - f(x)}{\Delta x}.$$

Note that we can take the derivative of a function (whatever that means) at particular values of x. For a point $c \in (a, b)$ (the interval on which $f(x)$ is differentiable), the derivative of $f(x)$ at $x = c$ follows easily from the definition:

$$\left.\frac{d}{dx}f(x)\right|_{x=c} = f'(c) = \lim_{\Delta x \to 0} \frac{f(c + \Delta x) - f(c)}{\Delta x}.$$

To understand the meaning of the derivative, we will reconcile it with the naïve equation for the slope of a line:

$$\text{slope} = \frac{\text{rise}}{\text{run}}.$$

What we mean by "rise" and "run" is the net vertical distance a function travels (rise) between the endpoints on some x-interval (run). Using this, a more precise formula for the slope (or steepness) of the function $f(x)$ over the interval $[a, b]$ is

$$\text{slope} = \frac{f(b) - f(a)}{b - a}.$$

Here, of course, $b > a$—suppose by some amount Δx. Then $b = a + \Delta x$[†] and so

$$\text{slope} = \frac{f(a + \Delta x) - f(a)}{(a + \Delta x) - a}$$
$$= \frac{f(a + \Delta x) - f(a)}{\Delta x}.$$

But this is the formula (minus a limit sign) for the derivative at $x = a$. Thus, we have shown that the slope of a function on some interval

---

[†]The reason we use Δx is because $b - a = \Delta x$ is the change in x over the interval $[a, b]$.

is intimately related to the derivative. If we now let $\Delta x \to 0$, what we find is an expression for the slope of the function $f(x)$ at the particular point $x = a$. In other words, it is the slope of the line tangent to the curve $f(x)$ at $x = a$. And this definition generalizes to the derivative at any point x—it is the meaning of the function $f'(x)$, the derivative of $f(x)$.

An identical interpretation of the derivative is that it measures the instantaneous rate of change of a function. That is, the derivative tells us by how much a function is increasing or decreasing at a particular point. This is a useful interpretation because it pours us into a subfield of calculus that deals with optimizations. One could ask: At what value of x in the interval $[a, b]$ is the function $f(x)$ maximized? One way to do this would be to graph $f(x)$ and look for the greatest peak in the interval. Now any mountaineer will tell you that a peak of a mountain is locally flat. And this is true of the peaks of functions—they are locally flat at their crests (and also their troughs). This is to say a function's instantaneous rate of change (i.e. derivative) is zero at its peaks and valleys. This means that whenever the derivative of a function vanishes, the function is either locally maximized or minimized at that point. We will see an example of this application of the derivative in §2.6.

We are now ready for everything that follows in this chapter. We begin with a few interesting applications of limits, then slowly transition to the land of differential calculus where we prove the optimal betting strategy in both fair and unfair coin-flipping games.

§2.1 Antipodal Temperatures and Wobbly Tables

Many foundational theorems in mathematics are intuitively obvious but their proofs quite involved. Such is the case with the *Intermediate Value Theorem* (IVT), which says that a continuous function $f(x)$ on $[a, b]$ will take on every value between $f(b)$ and $f(a)$. In

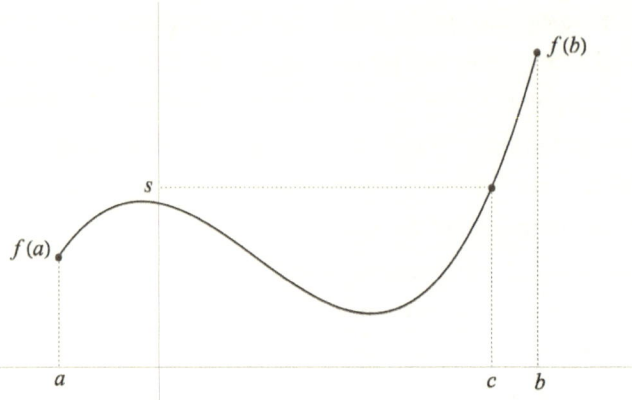

Figure 2.2: Since $f(a) < s < f(b)$, the IVT assures a $c \in (a,b)$ such that $f(c) = s$.

more rigorous terms, for a continuous function $f(x)$ defined on the interval $[a,b]$, for any $s \in \mathbb{R}$ such that $f(a) < s < f(b)$, there exists a $c \in (a,b)$ such that $f(c) = s$. Figure 2.2 sheds some light on the theorem—saying that because $f(b) > f(a)$, every value s in between $f(b)$ and $f(a)$ must have a corresponding x-value c such that $f(c) = s$. Though instinctive, the IVT has quite an involved proof. For sake of both perspicuity and brevity, we omit its proof and instead take the theorem for granted.

To make use of the theorem we will explore two problems that may not necessarily agree with your intuition. The first concerns the temperature at *antipodal points* on the surface of Earth—two points on Earth's surface that are directly opposite to one another (e.g. the North and South Poles.) Using the definition of a continuous function, we will prove that there always exists antipodal points on Earth's surface such that both are of equal temperature. In the second problem we intend to prove that square, four-legged wobbly tables can be made stable by at most a $\frac{\pi}{2}$ radian rotation. This, we hope, will be useful in impressing that special someone.

§2.1. ANTIPODAL TEMPERATURES AND WOBBLY TABLES   25

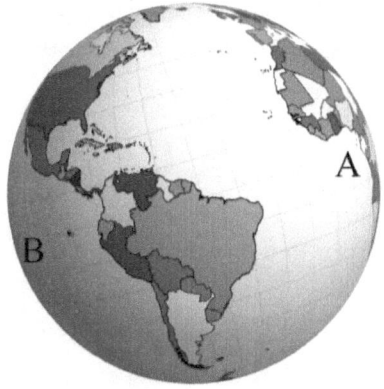

Figure 2.3: The Earth with persons A and B at antipodal points.

For the first, we require a function which monitors the temperature around the globe. Without loss of generality, we will prove the case on the Equator of the Earth, but realize the same argument extends to all circumscribable circles around the Earth. Define

$$T(l_o) := \text{the temperature on the Equator at longitude } l_o. \quad (2.4)$$

Since the Equator is (approximately) a circle, the interval on which l_o is defined is $[0, 2\pi)$ because $0 \leq l_0 < 2\pi$ (beyond this l_o is periodic). To illustrate (2.4), writing $T(0) = 79°\text{F}$ indicates the temperature on the Equator at longitude $l_o = 0$ radians (this is the location where the Equator crosses the Prime Meridian—it is in the Gulf of Guinea and is known as *Null Island*) is 79°F.[†]

Because the temperature on the surface of the Earth will not contain spots such that the temperature changes from, say, 60°F to 61°F without first jumping to an intermediate temperature of, say, 60.1°F, then 60.2°F, etc. (the laws of thermodynamics, to a certain extent, forbid this), it is fair to assume $T(l_o)$ is continuous for all

---

[†]This is the current temperature on 23 June 2017—the day we are writing this.

$l_o \in [0, 2\pi)$. Consequently, (2.4) is a continuous function.

To make progress on the proof, consider two people standing on the Equator at antipodal points, as in Figure 2.3. If person A is at longitude $l_o = \alpha \leq \pi$, then, due to their antipodal relationship, it must be that person B is at longitude $l_o = \pi + \alpha$. The problem requires us to find some α such that

$$T(\alpha) = T(\pi + \alpha) \implies T(\alpha) - T(\pi + \alpha) = 0°\text{F}.$$

In anticipation of the future analysis, it is useful to define a new function $\Psi(\alpha)$ as the difference in temperature between antipodal points on the Equator at longitudes $l_o = \alpha$ and $l_o = \pi + \alpha$. That is,

$$\Psi(\alpha) := T(\alpha) - T(\pi + \alpha).$$

Our interest is thus in α such that $\Psi(\alpha) = 0°\text{F}$.

Now suppose we position person A at Null Island—making $l_o = \alpha = 0$ by definition. Person B is then at longitude $l_0 = \pi$ and so the temperature difference is

$$\Psi(0) = T(0) - T(\pi).$$

If at these points $\Psi(0) = 0°\text{F}$, then we have found two antipodal points at the same temperature and so we are done. Suppose, however, $\Psi(0) \neq 0°\text{F}$. We then command both persons A and B to walk (or swim) at an identical pace along the Equator until person A is at longitude $l_o = \pi$ and person B is at longitude $l_o = 2\pi \implies l_o = 0.$[†] In other words, we have persons A and B travel at equal speed until they have circumnavigated exactly half the Earth across the Equator so that $\alpha = \pi$. Assuming temperatures did not change during the expedition (they walked and swam at an unimaginable speed), the temperature difference will be

$$\Psi(\pi) = T(\pi) - T(0) = -\Psi(0).$$

[†]This implication is due to the periodic nature of longitude, which is true since Earth is globular.

§2.1. Antipodal Temperatures and Wobbly Tables

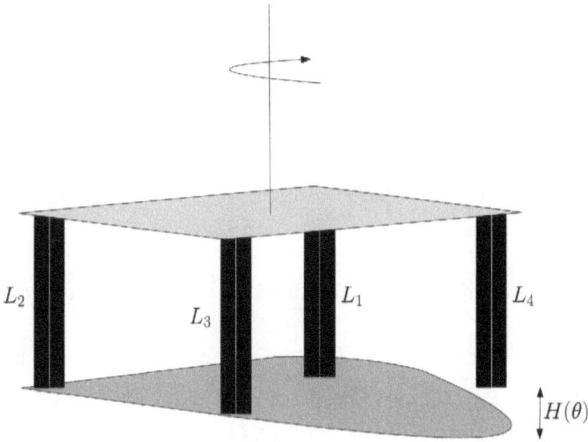

Figure 2.4: A rigid table with equal length legs L_1, L_2, L_3, and L_4. Here, L_1, L_2, and L_3 are positioned firmly against the ground, while L_4 is perched a height $H(\theta)$ above the ground, where θ is the angle through which the table is rotated.

Because we assumed $\Psi(0) \neq 0$, we require either $\Psi(\pi) > 0$ and $\Psi(0) < 0$ or $\Psi(\pi) < 0$ and $\Psi(0) > 0$. In either case, the IVT guarantees that for some $\alpha = \beta$, where $0 < \beta < \pi$, $\Psi(\beta) = 0°F$ because $\Psi(\alpha)$ is a continuous function.[†] Hence, when person A reaches longitude $l_o = \beta$ and person B longitude $l_o = \pi + \beta$, they will be at antipodal points whose temperature is the same.

What is incredible about this proof is that we need not consider temperature. The same result holds for altitude, pressure, luminosity, etc.—any physical measurement that can be modeled by a continuous function.

Let's now consider the second application of the IVT—showing that any square, four-legged wobbly table can be made stable with

---

[†]Indeed, the sum (or difference) of two continuous functions, such as $T(\alpha)$ and $T(\pi + \alpha)$, is itself a continuous function. Hence, $\Psi(\alpha)$ is continuous.

at most a quarter turn.† To prove this, consider Figure 2.4. Here, L_1 is the first leg, L_2 the second leg, and so forth, all of which are assumed to be of equal length. Without loss of generality, suppose the table can be pressed on so that legs L_1, L_2, and L_3 are placed on the ground, but that leg L_4 resides a height $H_0 > 0$ above the ground. We define

$H(\theta) :=$ height of L_4 after the table is rotated through an angle θ.‡

Obviously, of interest is $\theta \in [0, \frac{\pi}{2}]$. Beyond this, the table's position is periodic as L_1 assumes the place of L_2, L_2 of L_3, and so forth (remember, the table is square and so exhibits rotational symmetry). Observe with $\theta = 0$, $H(0) = H_0$.

We must prove that for some $\theta \in [0, \frac{\pi}{2}]$, $H(\theta) = 0$. Evidently, if $H(0) = H_0 = 0$, then the theorem is proven and the table is stable (again, we required all L_1, L_2, and L_3 to be secured to the ground). But suppose $H_0 \neq 0$. We'll then rotate the table slowly while keeping legs L_1, L_2, and L_3 pressed on the ground. If we rotate through a full $\frac{\pi}{2}$ radians, the legs take on the intransitive property

$$L_1 \to L_2 \to L_3 \to L_4 \to L_1 \tag{2.5}$$

as mentioned previously. In other words, all legs switch positions.

Plainly, if we perform this full rotation while constantly forcing L_1, L_2, and L_3 to be on the ground, L_4 will transition from height H_0 to some other height $H(\frac{\pi}{2})$, where $H(\frac{\pi}{2}) \neq 0$ due to the table's rotational symmetry. Yet upon rotating the table $\frac{\pi}{2}$ radians with legs L_1 through L_3 planted firmly on the ground, the only possibility is that $H(\frac{\pi}{2}) < 0$—that L_4 embeds itself into the ground. To see this, we exploit the intransitive property in (2.5). A $\frac{\pi}{2}$ rotation is the same as swapping all the legs. So the rotation is equivalent to legs L_2, L_3

---

†Our analysis assumes the insecurity of the table is derived from a rugged floor and not a defect in the table itself.
‡This height is measured relative to the ground.

§2.2. Euler's Number

and L_4 being against the ground and L_1 being the leg whose height is measured. But by the way the ground slopes down relative to L_4 before the rotation, pressing L_4 against the ground requires L_1 to go beneath the ground (see Figure 2.4 to build some intuition why this is necessarily the case). Hence, $H(\frac{\pi}{2}) < 0$.

Altogether, we have shown that for a $\frac{\pi}{2}$ rotation, the height function $H(\theta)$ proceeds from $H(0) = H_0 > 0$ to $H(\frac{\pi}{2}) < 0$. Since $H(\theta)$ is a continuous function on $[0, \frac{\pi}{2}]$, by the IVT there must exist some $\phi \in (0, \frac{\pi}{2})$ such that $H(\phi) = 0$. Such a ϕ is the angle through which the table must be rotated to make it stable.

What we've just proved is known as the *wobbly table theorem*. Perchance it will be of use the next time an unsteady table causes your beverage to tumble to the ground or, God forbid, spill on your date.[†]

---

§2.2 Euler's Number

We introduced the derivative $\frac{d}{dx}f(x)$ as the rate of change of the original function $f(x)$. In this section we intend to explore this definition and pursue the question:

> Is there a function whose rate of change function, i.e. derivative, is itself?

A natural approach to this question is to recall the definition of the derivative for a function $f(x)$:

$$\frac{d}{dx}f(x) := \lim_{\Delta x \to 0} \frac{f(x + \Delta x) - f(x)}{\Delta x}.$$

We are interested in a function $f(x)$ such that

$$\frac{d}{dx}f(x) = f(x) \implies \lim_{\Delta x \to 0} \frac{f(x + \Delta x) - f(x)}{\Delta x} = f(x).$$

[†] A minor concession: While the table will become stable, there is no guarantee it will be level. Nevertheless, this is a small price to pay for the security of a meal.

Though not immediately obvious, the best choice of $f(x)$ is an exponential function of the form a^x. To illustrate why, simply substitute a^x into the definition of the derivative:

$$\frac{d}{dx}a^x = \lim_{\Delta x \to 0} \frac{a^{x+\Delta x} - a^x}{\Delta x}.$$

By the exponent property $a^{b+c} = a^b \times a^c$, we have

$$\frac{d}{dx}a^x = \lim_{\Delta x \to 0} \frac{a^x \times a^{\Delta x} - a^x}{\Delta x} = \lim_{\Delta x \to 0} a^x \left(\frac{a^{\Delta x} - 1}{\Delta x}\right).$$

And because the function a^x has no dependence on Δx, we can bring it outside the limit sign to deduce

$$\frac{d}{dx}a^x = a^x \lim_{\Delta x \to 0} \left(\frac{a^{\Delta x} - 1}{\Delta x}\right). \tag{2.6}$$

Observe the lack of dependence on x in the limit. Assuming it exists, this absence by x implies the limit

$$\lim_{\Delta x \to 0} \left(\frac{a^{\Delta x} - 1}{\Delta x}\right)$$

is a constant. Hence, we have proved that the derivative of any general exponential function is proportional to itself. In notation, we write

$$\frac{d}{dx}a^x \propto a^x,$$

where $\propto$ means "proportional to."

As satisfying as this may seem, we have not yet answered the question, though we've made significant progress. We have reduced our search from a general function $f(x)$ to an exponential function a^x such that

$$\frac{d}{dx}a^x = a^x. \tag{2.7}$$

For notational (and historic) purposes, we redefine the unique base with the property in (2.7) as e instead of a. Hence, we are in search for the value of e such that

$$\frac{d}{dx}e^x = e^x.$$

§2.2. Euler's Number

By (2.6), this is equivalent to finding the value of e such that the limit

$$\lim_{\Delta x \to 0} \left(\frac{e^{\Delta x} - 1}{\Delta x} \right) = 1. \tag{2.8}$$

Provided we are careful with the limit sign, we can rearrange (2.8) to get e by itself. Multiplying both sides both Δx (and carrying over the limit), we have

$$\lim_{\Delta x \to 0} e^{\Delta x} - 1 = \lim_{\Delta x \to 0} \Delta x.$$

We then add one to both sides:

$$\lim_{\Delta x \to 0} e^{\Delta x} = \lim_{\Delta x \to 0} 1 + \Delta x.$$

Finally, we raise both sides to the $\frac{1}{\Delta x}$ power:

$$\left(\lim_{\Delta x \to 0} e^{\Delta x} \right)^{\frac{1}{\Delta x}} = \left(\lim_{\Delta x \to 0} 1 + \Delta x \right)^{\frac{1}{\Delta x}}$$

$$\lim_{\Delta x \to 0} \left(e^{\Delta x} \right)^{\frac{1}{\Delta x}} = \lim_{\Delta x \to 0} (1 + \Delta x)^{\frac{1}{\Delta x}}$$

$$\lim_{\Delta x \to 0} e = \lim_{\Delta x \to 0} (1 + \Delta x)^{\frac{1}{\Delta x}}$$

Here, the limit on the left hand side is just e because e is a constant. Thus,

$$e = \lim_{\Delta x \to 0} (1 + \Delta x)^{\frac{1}{\Delta x}}.$$

To go one step further, we will introduce the substitution $n = \frac{1}{\Delta x}$, which is justified provided we change the limit $\Delta x \to 0$ to $n \to \infty$. Substituting this in, we obtain

$$e := \lim_{n \to \infty} \left(1 + \frac{1}{n} \right)^n. \tag{2.9}$$

Our use of $:=$ indicates that this is indeed the definition of the value e, which is often called *Euler's number* after the great eighteenth century mathematician Leonhard Euler.

To get a quantitative idea on the value of e, we can substitute various n into (2.9). This generates the table

n	$\left(1+\frac{1}{n}\right)^n$
1	2
10	$2.593742460100002\cdots$
100	$2.704813829421528\cdots$
10000	$2.718145926824925\cdots$
10000000	$2.718281694132081\cdots$
∞	$2.718281828459045\cdots$

Consequently,

$$e = 2.718281828459045\cdots.$$

The use of ellipsis is intentional because e is known to be irrational (this we prove in §4.6) so its decimal expansion is infinite.

We will see many applications and problems concerned with Euler's number in future sections. Hopefully by the end of this book it will be evident that e is quite an important number, comparable to the likes of π.

§2.3 An Arduous Limit

Often the evaluation of limits is systematic and straightforward. Only in certain cases do limits appear reluctant in shedding their value, and yet even here the procedure for calculating the limit is probably nothing special. This section is devoted to a particularly cumbersome limit whose evaluation may initially appear painless, but whose proper justification is quite involved (but fun nonetheless). The limit in question is

$$\lim_{\phi \to 0} \frac{\sin(\phi)}{\phi}. \tag{2.10}$$

This is the quintessential limit that many try and solve using an extraordinary theorem called *L'Hôpital's rule*, named after the seventeenth century mathematician Guillaume de L'Hôpital. We will

§2.3. An Arduous Limit

shortly demonstrate that using this rule to determine (2.10) is actually a circular argument, and so is an illogical approach. Of course, before doing this we require an understanding of what L'Hôpital's rule actually says.

Given two continuous functions $f(x)$ and $g(x)$, L'Hôpital's rule states that if the limit
$$\lim_{x \to a} \frac{f(x)}{g(x)}$$
approaches an indeterminate form such as $\pm \frac{\infty}{\infty}$ or $\frac{0}{0}$, then
$$\lim_{x \to a} \frac{f(x)}{g(x)} = \lim_{x \to a} \frac{f'(x)}{g'(x)},$$
where $f'(x)$ and $g'(x)$ are the derivatives of $f(x)$ and $g(x)$, respectively. To demonstrate the rule, consider the functions $f(x) := x^n$ and $g(x) := e^x$, where $n \in \mathbb{N}$. We are interested in the limiting behavior of the ratio $\frac{x^n}{e^x}$ as $x \to \infty$. Substituting in $x = \infty$, we obtain the indeterminate form
$$\frac{\infty^n}{e^\infty} = \frac{\infty}{\infty},$$
so L'Hôpital's rule is applicable. Differentiating both $f(x)$ and $g(x)$, we find $f'(x) = nx^{n-1}$ and $g'(x) = e^x$. Thus, by L'Hôpital's rule we have
$$\lim_{x \to \infty} \frac{x^n}{e^x} = \lim_{x \to \infty} \frac{nx^{n-1}}{e^x} = \frac{n\infty^{n-1}}{e^\infty} = \frac{\infty}{\infty}.$$
This does not seem to have helped us that much because we still have an indeterminate form. But this means L'Hôpital's rule applies yet again. So we can differentiate the top and bottom functions nx^{n-1} and e^x to deduce the limiting behavior of the ratio $\frac{nx^{n-1}}{e^x}$ as $x \to \infty$, which in turn equals the limiting behavior of the original ratio $\frac{x^n}{e^x}$. Since $\frac{d}{dx}nx^{n-1} = n(n-1)x^{n-2}$ and $\frac{d}{dx}e^x = e^x$, we have that
$$\lim_{x \to \infty} \frac{nx^{n-1}}{e^x} = \lim_{x \to \infty} \frac{n(n-1)x^{n-2}}{e^x} = \frac{n(n-1)\infty^{n-2}}{e^\infty} = \frac{\infty}{\infty}.$$

By now we recognize the pattern: Each iteration yields an indeterminate form, the top function decreases by a power of one, and the bottom function remains the same. Because n is a whole number, it can only decrease so much before reaching zero. At this point (a total of n iterations) we will obtain the limit

$$\lim_{x \to \infty} \frac{n \times (n-1) \times (n-2) \times \cdots \times 2 \times 1}{e^x}$$
$$= \frac{n \times (n-1) \times (n-2) \times \cdots \times 2 \times 1}{e^\infty}$$
$$= \frac{n \times (n-1) \times (n-2) \times \cdots \times 2 \times 1}{\infty}$$
$$= \frac{n!}{\infty}$$
$$\to 0,$$

where in the penultimate step we utilized the *factorial function*

$$n! := n \times (n-1) \times (n-2) \times \cdots \times 2 \times 1.$$

Because we just performed many applications of L'Hôpital's rule on the same two functions, we have the cascading equality

$$\lim_{x \to \infty} \frac{x^n}{e^x} = \lim_{x \to \infty} \frac{nx^{n-1}}{e^x} = \cdots = \lim_{x \to \infty} \frac{n!}{e^x} = 0.$$

Consequently, we have shown

$$\lim_{x \to \infty} \frac{x^n}{e^x} = 0,$$

which implies the exponential function e^x grows faster than every polynomial function of integer degree (indeed, this is true of every polynomial with a constant degree).

Back to the problem at hand, we are interested in the value to which the ratio $\frac{\sin(\phi)}{\phi}$ approaches as $\phi \to 0$ (this is (2.10)). Notice that by substituting $\phi = 0$ into (2.10), we obtain

$$\frac{\sin(0)}{0} = \frac{0}{0},$$

§2.3. An Arduous Limit

which is an indeterminate form, so L'Hôpital's rule is applicable. Since $\frac{d}{d\phi}\sin(\phi) = \cos(\phi)$ and $\frac{d}{d\phi}\phi = 1$, L'Hôpital's rule tells us that

$$\lim_{\phi \to 0} \frac{\sin(\phi)}{\phi} = \lim_{\phi \to 0} \frac{\cos(\phi)}{1} = 1. \tag{2.11}$$

Now all would be well if we remain ignorant to what the derivative actually is. But because we are rational and rigorous mathematicians, we will not do this.

The definition of the derivative for some function $f(x)$, as we've regurgitated many times now, is

$$\frac{d}{dx} f(x) := \lim_{\Delta x \to 0} \frac{f(x + \Delta x) - f(x)}{\Delta x}.$$

Hence, $\frac{d}{d\phi}\sin(\phi)$ is defined as the limit

$$\lim_{\Delta \phi \to 0} \frac{\sin(\phi + \Delta \phi) - \sin(\phi)}{\Delta \phi}. \tag{2.12}$$

To get a better handle on this, we utilize the trigonometric addition formula for the sine function:

$$\sin(A + B) = \sin(A)\cos(B) + \sin(B)\cos(A).^{\dagger}$$

Applying this to the term $\sin(\phi + \Delta\phi)$ in (2.12) generates

$$\frac{d}{d\phi}\sin(\phi) = \lim_{\Delta\phi \to 0} \frac{\sin(\phi)\cos(\Delta\phi) + \sin(\Delta\phi)\cos(\phi) - \sin(\phi)}{\Delta\phi}.$$

A simple algebraic rearrangement prompts

$$\frac{d}{d\phi}\sin(\phi)$$
$$= \sin(\phi) \left(\lim_{\Delta\phi \to 0} \frac{\cos(\Delta\phi) - 1}{\Delta\phi} \right) + \cos(\phi) \left(\lim_{\Delta\phi \to 0} \frac{\sin(\Delta\phi)}{\Delta\phi} \right). \tag{2.13}$$

---

†This is derived in §4.10.

Notice the second term of (2.13) is precisely the limit in (2.10). So in stating
$$\frac{d}{d\phi}\sin(\phi) = \cos(\phi)$$
(what we did in evaluating (2.10) with L'Hôpital's rule) we assume we know the value to which (2.10) converged. But this is an entirely circular justification for (2.11), in the same way the argument "I am right because I am right" is circular: You cannot prove something correct if your argument is contingent upon that thing being correct. Therefore, using L'Hôpital's rule to determine (2.10) is misguided.

This examples illustrates a more general and subtle point about mathematics: One must always proceed carefully through a problem and be aware of all underlying assumptions in every operation they perform, even if it is something as simple as a derivative. This rigor contributes to the general (though incomplete) explanation for why mathematics is such a successful discipline. And it is why many prescribe beauty to its consistency.

Now the question remains, how might we properly show that indeed $\frac{\sin(\phi)}{\phi} \to 1$ as $\phi \to 0$? Though this can be done using a myriad of techniques, the most elegant stems from the geometric construction in Figure 2.5. This graphic represents the first quadrant of the unit circle, meaning $\overline{WX} = \overline{WZ} = 1$ because the radius of the unit circle is one. If we denote the area of $\triangle WXZ$ by A_1, then by simple geometry in Figure 2.5 we find
$$A_1 = \frac{1}{2}(1)h = \frac{1}{2}\sin(\phi).$$
For $\triangle WYZ$, whose area we call A_2, observe that
$$\tan(\phi) = \frac{\overline{YZ}}{\overline{WZ}} = \frac{\overline{YZ}}{1} = \overline{YZ}.$$
Therefore, the area A_2 is
$$A_2 = \frac{1}{2}(1)\overline{YZ} = \frac{1}{2}\tan(\phi).$$

§2.3. An Arduous Limit

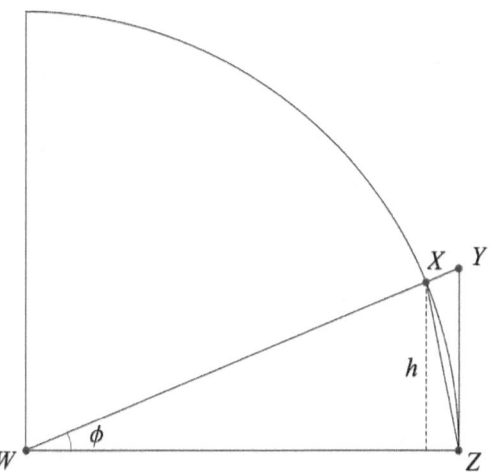

Figure 2.5: A portion of the unit circle. Thus, the radius $= \overline{WX} = \overline{WZ} = 1$.

Finally, because $\widehat{XZ}$ subtends the angle ϕ, the area A_3 of the sector bounded by $\widehat{XZ}$, $\overline{WX}$, and $\overline{WZ}$ is

$$A_3 = \frac{1}{2}\phi.$$

Examining Figure 2.5 it is clear that for $0 < \phi < \frac{\pi}{2}$,

$$A_1 \leq A_3 \leq A_2 \implies \frac{1}{2}\sin(\phi) \leq \frac{1}{2}\phi \leq \frac{1}{2}\tan(\phi).$$

Multiplying through by a factor of $\frac{2}{\sin(\phi)}$, we acquire the compound inequality

$$1 \leq \frac{\phi}{\sin(\phi)} \leq \frac{1}{\cos(\phi)}. \tag{2.14}$$

Taking the reciprocal of each element in (2.14), we obtain

$$1 \geq \frac{\sin(\phi)}{\phi} \geq \cos(\phi). \tag{2.15}$$

We are interested in the value to which the function $\frac{\sin(\phi)}{\phi}$ nears as $\phi \to 0$. Looking at (2.15), we see that for ϕ such that $0 < \phi < \frac{\pi}{2}$, the function $\frac{\sin(\phi)}{\phi}$ is always between one and $\cos(\phi)$. Letting $\phi \to 0$, (2.15) reduces to

$$\lim_{\phi \to 0} 1 \geq \lim_{\phi \to 0} \frac{\sin(\phi)}{\phi} \geq \lim_{\phi \to 0} \cos(\phi) \implies 1 \geq \lim_{\phi \to 0} \frac{\sin(\phi)}{\phi} \geq 1 \tag{2.16}$$

because

$$\lim_{\phi \to 0} \cos(\phi) = \cos(0) = 1.$$

Thus, it must be that

$$\lim_{\phi \to 0} \frac{\sin(\phi)}{\phi} = 1$$

because the ratio $\frac{\sin(\phi)}{\phi}$ is essentially squeezed between the two values of one in (2.16).[†]

While this method is certainly more involved than L'Hôpital's rule, at least it is genuine. One must never forget the power of geometry.

§2.4 The Dirichlet Function

In §1.1 we introduced the set $\mathbb{Q}$—the home for all the rational numbers. Obviously, $\mathbb{Q}$ accommodates many numbers (an infinite number, in fact), so keeping track of all the rationals can be a difficult task. A tenable question is thus: Given some arbitrary real number r, how might we determine if it is rational or not? One way to approach this would be to find two integers a and b such that the ratio $\frac{a}{b} = r$. This would prove $r \in \mathbb{Q}$. Though, there are so many possible combinations of a and b that this approach is monotonous

---

[†]Here the word "squeezed" is used intentionally since our final assertion follows from something called the *squeeze theorem*. This says that given three functions $f(x), g(x)$, and $h(x)$ such that for some value a, $\lim_{x \to a} f(x) = \lim_{x \to a} h(x) = L$ and

$$\lim_{x \to a} f(x) \leq \lim_{x \to a} g(x) \leq \lim_{x \to a} h(x),$$

then $g(x)$ necessarily approaches L as $x \to a$. That is, $\lim_{x \to a} g(x) = L$.

§2.4. The Dirichlet Function

and inefficient. What we explore in this section is a different, expedited, and overall ingenious method for determining whether a number is rational or not.

The analysis begins with the *Dirichlet function* ("dear-ish-lay"), named after the nineteenth century mathematician Peter Gustav Lejeune Dirichlet. His function is defined as the piecewise conglomeration

$$D(x) := \begin{cases} 1 & \text{if } x \in \mathbb{Q} \\ 0 & \text{if } x \notin \mathbb{Q} \end{cases} \tag{2.17}$$

for all $x \in \mathbb{R}$. In more heuristic terms, (2.17) translates to

$$D(\text{rational}) = 1 \text{ and } D(\text{irrational}) = 0.$$

This function, however, is quite futile. For starters, (2.17) provides no mechanism for determining whether a number is rational or not, it merely assigns a boolean to the input based on our understanding of its rationality (or lack thereof). What would be infinitely more useful is some function $\delta(x)$ which equals (2.17) and also provides a means for determining if the input is rational. We spend the remainder of this section deducing how the following choice of $\delta(x)$ does just this:

$$\delta(x) = \lim_{\substack{n \to \infty \\ k \to \infty}} \cos^{2n}(k!\pi x) = D(x).^{\dagger} \tag{2.18}$$

To understand how this works, we will break (2.18) into its main components and steadily piece them back together. We start with $\cos(\pi x)$.

As is evident from the trigonometric identity $\cos^2(x) + \sin^2(x) = 1$, whenever $\sin^2(x)$ vanishes, $\cos(x) = \pm 1$. Hence, the cosine

---

†The double limit notation

$$\lim_{\substack{n \to \infty \\ k \to \infty}} = \lim_{n \to \infty} \lim_{k \to \infty}$$

is the same as taking two limits simultaneously on a function.

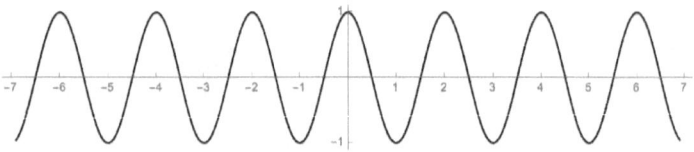

Figure 2.6: The graph of $\cos(\pi x)$. Note the extrema whenever $x \in \mathbb{Z}$.

function is bounded between -1 and 1 for all real x. The values of x at which $\cos(x) = \pm 1$ (i.e. the cosine function's extrema) are computed by setting the derivative of $\cos(x)$ equal to zero and solving for x:[†]

$$\frac{d}{dx}\cos(x) = 0 \implies \sin(x) = 0.$$

This, we know, is true for all x such that $x = k\pi$ where $k \in \mathbb{Z}$. Therefore, $\cos(x)$ is maximized and minimized at all values $x = k\pi$. This property is reflected in the graph of $\cos(\pi x)$ in Figure 2.6. All the extrema occur whenever x is an integer, as predicted. Moreover, note that the value of the extrema are all ± 1.

We claimed in (2.18) that $\delta(x) = D(x)$. This means $\delta(x)$ outputs a boolean in the same way $D(x)$ does. Hence, our reconstruction of $\delta(x)$ must output either zero or one. We can do this by using the extrema properties of $\cos(\pi x)$. By the analysis above, for any input $x \notin \mathbb{Z}$ the function $\cos(\pi x)$ is such that $-1 < \cos(\pi x) < 1$. Hence, we claim the limit

$$\lim_{n \to \infty} \cos^n(\pi x) = 0 \qquad (2.19)$$

[†]We mention why this is in the discussion towards the end of this chapter's introduction.

§2.4. The Dirichlet Function

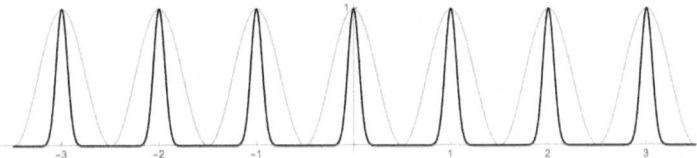

Figure 2.7: The graphs of $\cos^2(\pi x)$ (light line) and $\cos^{42}(\pi x)$ (dark line).

for all $x \notin \mathbb{Z}$.[†] It will be useful later on to utilize the identical result

$$\lim_{n \to \infty} (\cos^n(\pi x))^2 = 0^2 = 0$$
$$\implies \lim_{n \to \infty} \cos^{2n}(\pi x) = 0 \qquad (2.20)$$

so that for all possible x and n, the limit approaches zero from positive numbers (or, in the case $x \in \mathbb{Z}$, the limit approaches *only* positive one). This way we need not concern ourselves with negative values.

Figure 2.7 portrays what (2.20) is doing. Notice for $\cos^2(\pi x)$ the extrema are all positive (hence the desire to square $\cos(\pi x)$ in (2.20)). The same is true for $\cos^{42}(\pi x)$, except the peaks are steeper with elongated tails. As one might guess, letting $n \to \infty$ as in (2.20) makes all non-integer x converge to zero while those which are integers remain at unity. In all, we have constructed a function

---

[†] (2.19) is characteristic of such an omnipotent limit that we take a small detour to talk about it. Consider the limit $\lim_{n \to \infty} x^n$. We ask: For what $x \in \mathbb{R}$ does this converge and for what x does it diverge? Notice for $|x| > 1$, the limit just multiplies together an infinite string of x's, getting larger and larger since $|x| > 1$. Hence, it diverges. For $x = 1$, the limit is a product of ones, so the limit is one. Finally, for $|x| < 1$, the limit is a product of fractions (or irrational numbers) in between -1 and 1. Each successive product is smaller than the previous, so the limit approaches zero, as in (2.19).

$\rho(x)$ that tells us whether a number is an integer or not:

$$\rho(x) = \lim_{n \to \infty} \cos^{2n}(\pi x) = \begin{cases} 1 & \text{if } x \in \mathbb{Z} \\ 0 & \text{if } x \notin \mathbb{Z}. \end{cases} \quad (2.21)$$

While similar in form to (2.17), we must push a little further to obtain $D(x)$ entirely.

What we would like to do now is construct some function $\tau(x)$ that can distinguish between rational and irrational inputs. A particularly useful $\tau(x)$ would be one that maps rationals to integers and irrationals to non-integers. This way, applying (2.21) to the output of $\tau(x)$ would yield one for rational x and zero for irrational x. Overall, rationals go to integers and thus output one, while irrationals go to non-integers and so output zero. Symbolically, once $\tau(x)$ is obtained, $\delta(x)$ will be the composite function

$$\delta(x) = (\rho \circ \tau)(x) = \rho(\tau(x)) = D(x). \quad (2.22)$$

To construct such a rational-to-integer function $\tau(x)$, it is best to restate the definition of a rational number: A number q is rational if it can be expressed as the ratio $\frac{a}{b}$ where $a, b \in \mathbb{Z}$ and $b \neq 0$. One way to convert q to an integer is to multiply it by b so that

$$q \times b = \frac{a}{b} \times b = a \in \mathbb{Z}.$$

Of course, this method necessitates an understanding of q being rational because we need to know the denominator b in q's integer ratio. One way to circumvent this complication is to multiply q by the limit

$$\lim_{k \to \infty} k!, \quad (2.23)$$

where $k! := k \times (k-1) \times \cdots \times 2 \times 1$ is the factorial function. This way, no matter the magnitude of b, the product

$$q \times \lim_{k \to \infty} k! = \frac{a}{b} \times \lim_{k \to \infty} k!$$

§2.4. THE DIRICHLET FUNCTION

will eventually surpass b and algebraically cancel it, leaving a string of integers that are multiplied together:

$$\frac{a}{b} \times \lim_{k \to \infty} k! = a \times 1 \times 2 \times \cdots \times (b-1) \times (b+1) \times \cdots.$$

This cancellation, however, will only occur provided $q \in \mathbb{Q}$. If q is not expressible as the ratio $\frac{a}{b}$ ($q \notin \mathbb{Q}$), then no matter how large k becomes in (2.23), $q \times k!$ will never be brought to an integer value. We therefore have the rational-to-integer function $\tau(x)$ we've been searching for:

$$\tau(x) = x \lim_{k \to \infty} k!. \tag{2.24}$$

To be explicit about its computation, we have determined that (2.24) holds the following property:

$$\begin{cases} \tau(x) \in \mathbb{Z} & \text{if } x \in \mathbb{Q} \\ \tau(x) \notin \mathbb{Z} & \text{if } x \notin \mathbb{Q}. \end{cases}$$

In other words,

$$\tau(\text{rational}) \to \text{integer}$$

while

$$\tau(\text{irrational}) \to \text{non-integer}.$$

Substituting (2.24) into (2.22) constructs a formula for $\delta(x)$:

$$\delta(x) = \rho(\tau(x))$$
$$= \rho\left(x \lim_{k \to \infty} k!\right)$$
$$= \lim_{n \to \infty} \cos^{2n}\left(\pi x \lim_{k \to \infty} k!\right).$$

Because k only appears in the argument of the cosine function, we can bring the limit outside. Therefore,

$$\delta(x) = \lim_{n \to \infty} \lim_{k \to \infty} \cos^{2n}(k!\pi x) = \lim_{\substack{n \to \infty \\ k \to \infty}} \cos^{2n}(k!\pi x),$$

which proves

$$\lim_{\substack{n \to \infty \\ k \to \infty}} \cos^{2n}(k!\pi x) = \begin{cases} 1 & \text{if } x \in \mathbb{Q} \\ 0 & \text{if } x \notin \mathbb{Q} \end{cases}$$

as claimed in (2.18). In other words, the function $\delta(x)$ is one which determines whether a given real number is rational or not—a fascinating property, to say the least.

§2.5 The Koch Snowflake

Perhaps the most pure form of mathematical intuition stems from geometry. Being able to see beyond an arcane equation to its underlying geometric meaning only provides deeper insight into the mechanics of that equation. But, contrary to how many perceive the field, geometry is not always unambiguous and intuitive. Indeed, geometry itself encompasses many objects whose properties appear to defy common sense. In this section, we investigate one such obscure object whose perimeter is infinite but area is finite. A physical consequence of this outstandingly outré combination is the ability to shade in the shape with a pencil, but an inability to trace it out. Many geometric shapes with this property are called *fractals*, and the one we explore here is the *Koch snowflake* discovered by the nineteenth-twentieth century mathematician Helge von Koch.

The construction of the Koch snowflake is quite simple. As you can see in Figure 2.8, the assembly begins with an equilateral triangle. We then divide each side length into a third, which forms the side length of a new equilateral triangle. We place this new triangle on the edge of the previous triangle such that the new triangle's center lies directly above the midpoint of each edge on the previous triangle. Repeating this process generates the recursive operation seen in Figure 2.8.

If n denotes the number of recursive iterations, defined such that at $n = 0$ we have the equilateral triangle in Figure 2.8, then the

§2.5. The Koch Snowflake

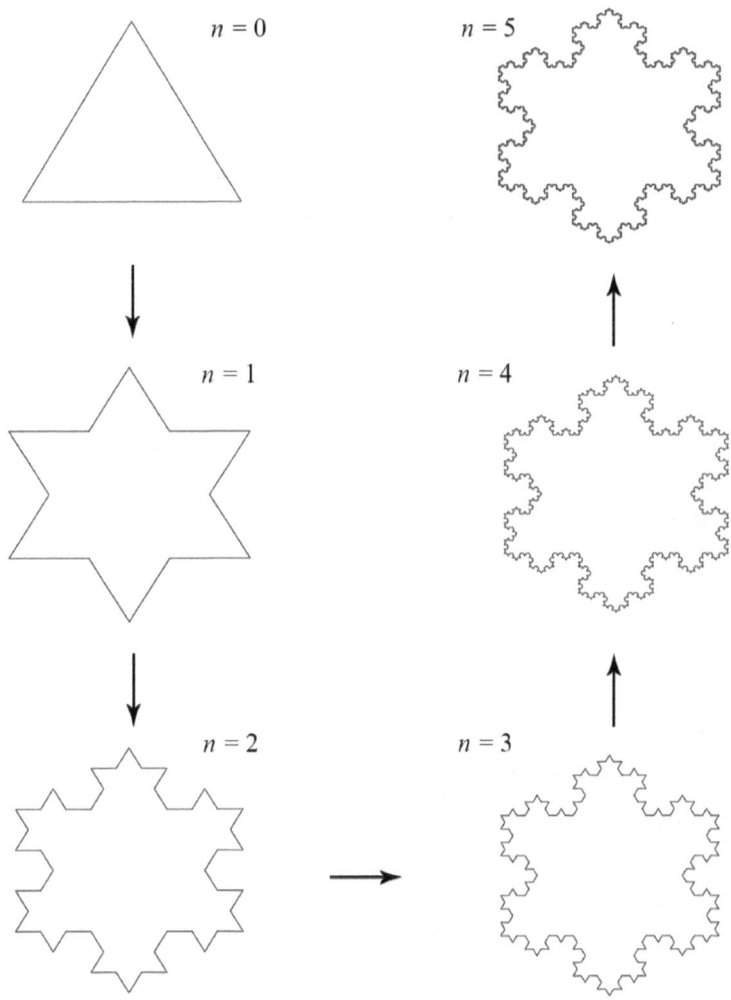

Figure 2.8: The construction of the Koch snowflake at various iterations n.

Koch snowflake is obtained in the limit as $n \to \infty$. Though impossible to illustrate its complexity on paper (ink can only provide so much detail), we can get an idea for the intricate nature of the Koch snowflake by computing its area and perimeter (and by marveling at the fifth iteration in Figure 2.8).

We'll start with area. Let s be the side length of the equilateral triangle and let A_n denote the area of the shape on the nth iteration. It follows from basic trigonometry that at $n = 0$ (the zeroth iteration) the area is

$$A_0 = \frac{1}{2}bh = \frac{1}{2}s\left(s\frac{\sqrt{3}}{2}\right) = \frac{\sqrt{3}}{4}s^2 \qquad (2.25)$$

since an equilateral triangle has $b = s$ and equal angles of $\frac{\pi}{3}$ radians, making $h = s\sin(\frac{\pi}{3}) = s\frac{\sqrt{3}}{2}$. For the next iteration ($n = 1$), we append three identical equilateral triangles with side length $\frac{1}{3}s$. Hence, the total area of the figure is

$$A_1 = A_0 + 3\overbrace{\left(\frac{\sqrt{3}}{4}\left(\frac{s}{3}\right)^2\right)}^{\text{area of one new triangle}}$$
$$= A_0 + \frac{3}{4} \times 4\left(\frac{\sqrt{3}}{4}\left(\frac{s}{3}\right)^2\right).^\dagger$$

Rather than simplify this expression, we leave it as is in hopes to deduce a general *recurrence relationship* between A_{n+1} and A_n.

For the next iteration ($n = 2$), we see from Figure 2.8 that twelve additional equilateral triangles are appended, each with side length $\frac{1}{3}\left(\frac{1}{3}s\right) = \frac{1}{3^2}s$. Therefore, the total area of the figure at $n = 2$ is

$$A_2 = A_1 + 12\left(\frac{\sqrt{3}}{4}\left(\frac{s}{3^2}\right)^2\right)$$
$$= A_1 + \frac{3}{4} \times 4^2\left(\frac{\sqrt{3}}{4}\left(\frac{s}{3^2}\right)^2\right).$$

---

†The additional factor of $\frac{1}{4}$ will allow for a more fruitful pattern in the near future.

§2.5. The Koch Snowflake

Finally, we count from Figure 2.8 that one more iteration ($n = 3$) yields forty-eight additional equilateral triangles, each with side length $\frac{1}{3}\left(\frac{1}{3^2}s\right) = \frac{1}{3^3}s$. It follows that the area is

$$A_3 = A_2 + 48\left(\frac{\sqrt{3}}{4}\left(\frac{s}{3^3}\right)^2\right)$$
$$= A_2 + \frac{3}{4} \times 4^3 \left(\frac{\sqrt{3}}{4}\left(\frac{s}{3^3}\right)^2\right).$$

In general, we postulate that the area after the $(n+1)$th iteration is

$$A_{n+1} = A_n + \frac{3}{4} \times 4^{n+1} \left(\frac{\sqrt{3}}{4}\left(\frac{s}{3^{n+1}}\right)^2\right), \qquad (2.26)$$

which is the recurrence relationship we've been searching for. To properly solve (2.26) for A_{n+1} in terms of only n (that is, devoid of terms like A_n, A_{n-1}, etc.), it is common to "unroll" the recurrence relationship and, again, deduce the general pattern.

It follows from (2.26) that

$$A_n = A_{n-1} + \frac{3}{4} \times 4^n \left(\frac{\sqrt{3}}{4}\left(\frac{s}{3^n}\right)^2\right),$$

making

$$A_{n+1} = A_{n-1} + \frac{3}{4} \times 4^n \left(\frac{\sqrt{3}}{4}\left(\frac{s}{3^n}\right)^2\right)$$
$$+ \frac{3}{4} \times 4^{n+1} \left(\frac{\sqrt{3}}{4}\left(\frac{s}{3^{n+1}}\right)^2\right)$$
$$= A_{n-1} + \frac{3\sqrt{3}}{16}s^2 \left(\frac{4^n}{3^{2n}} + \frac{4^{n+1}}{3^{2(n+1)}}\right).$$

Again, from (2.26) we know

$$A_{n-1} = A_{n-2} + \frac{3}{4} \times 4^{n-1} \left(\frac{\sqrt{3}}{4}\left(\frac{s}{3^{n-1}}\right)^2\right),$$

from which A_{n+1} becomes

$$A_{n+1} = A_{n-2} + \frac{3}{4} \times 4^{n-1} \left(\frac{\sqrt{3}}{4} \left(\frac{s}{3^{n-1}} \right)^2 \right)$$
$$+ \frac{3\sqrt{3}}{16} s^2 \left(\frac{4^n}{3^{2n}} + \frac{4^{n+1}}{3^{2(n+1)}} \right)$$
$$= A_{n-2} + \frac{3\sqrt{3}}{16} s^2 \left(\frac{4^{n-1}}{3^{2(n-1)}} + \frac{4^n}{3^{2n}} + \frac{4^{n+1}}{3^{2(n+1)}} \right).$$

In general, it appears A_{n+1} is governed by

$$A_{n+1} = A_0 + \frac{3\sqrt{3}}{16} s^2 \left(\frac{4}{3^2} + \frac{4^2}{3^4} + \frac{4^3}{3^6} + \cdots + \frac{4^{n+1}}{3^{2(n+1)}} \right), \quad (2.27)$$

where A_0 is provided in (2.25).

The sum on the right of (2.27) is itself a recurrence relationship. It is shown in §A.2 that any sum of the form

$$C + C\omega + C\omega^2 + \cdots + C\omega^n = \frac{C(1 - \omega^{n+1})}{1 - \omega}, \quad (2.28)$$

provided C and ω are constants. Relating this sum to the one in (2.27), it's fairly evident that

$$\frac{4}{3^2} + \frac{4^2}{3^4} + \frac{4^3}{3^6} + \cdots + \frac{4^{n+1}}{3^{2(n+1)}}$$
$$= \frac{4}{3^2} + \frac{4}{3^2} \left(\frac{4}{3^2} \right) + \frac{4}{3^2} \left(\frac{4}{3^2} \right)^2 + \cdots + \frac{4}{3^2} \left(\frac{4}{3^2} \right)^n$$

making $C = \omega = \frac{4}{3^2} = \frac{4}{9}$. Hence, it follows from (2.28) that

$$\frac{4}{3^2} + \frac{4^2}{3^4} + \frac{4^3}{3^6} + \cdots + \frac{4^{n+1}}{3^{2(n+1)}} = \frac{4}{9} \left[\frac{1 - \left(\frac{4}{9} \right)^{n+1}}{1 - \frac{4}{9}} \right]$$
$$= \frac{4}{5} \left[1 - \left(\frac{4}{9} \right)^{n+1} \right].$$

Therefore, after substituting in A_0 from (2.25) into (2.27) and using the result directly above, we have

$$A_{n+1} = \frac{\sqrt{3}}{4} s^2 + \frac{3\sqrt{3}}{20} s^2 \left[1 - \left(\frac{4}{9} \right)^{n+1} \right].$$

§2.5. THE KOCH SNOWFLAKE

The Koch snowflake is obtained in the limit as $n \to \infty$, thus the area of the Koch snowflake is given by the limit

$$\lim_{n \to \infty} A_{n+1} = \lim_{n \to \infty} \frac{\sqrt{3}}{4}s^2 + \frac{3\sqrt{3}}{20}s^2 \left[1 - \left(\frac{4}{9}\right)^{n+1}\right].$$

Since $\left|\frac{4}{9}\right| < 1$, we have that as $n \to \infty$, $\left(\frac{4}{9}\right)^{n+1} \to 0$.[†] Consequently, denoting the area of the Koch snowflake by A_∞, we have

$$A_\infty = \lim_{n \to \infty} A_{n+1} = \frac{\sqrt{3}}{4}s^2 + \frac{3\sqrt{3}}{20}s^2(1 - 0).$$

This simplifies to

$$A_\infty = \frac{2\sqrt{3}}{5}s^2, \qquad (2.29)$$

which is both a finite number and, curiously, $\frac{8}{5}A_0$.

This required quite a lot of algebra, but the trudge was ultimately worth it. We have shown the Koch snowflake has a *finite* area given by (2.29). We'll now compute the snowflake's perimeter.

Let p_n denote the perimeter of the Koch snowflake after the nth iteration. We define $p_0 = 3s$ as the perimeter of the equilateral triangle ($n = 0$) in Figure 2.8. At the next iteration ($n = 1$), three triangles are added—a total of $3 \times 3 = 9$ new edges, each with length $\frac{1}{3}s$. Note, however, that for every triangle we add, one edge is planted on the border of the previous $(n-1)$th snowflake. So to obtain only the outside perimeter, we must subtract this and all similar superfluous edges. As a result, the perimeter for $n = 1$ is

$$p_1 = p_0 + \overbrace{3 \times 3\left(\frac{s}{3}\right)}^{\text{total additional edges}} - \overbrace{3\left(\frac{s}{3}\right)}^{\text{superfluous edges}}$$

$$= p_0 + 2^2 \times \frac{3}{2}\left(\frac{s}{3}\right).[‡]$$

Playing the same game for $n = 2$, we add an additional twelve triangles for a total of $3 \times 12 = 36$ new edges, each with length $\frac{1}{3^2}s$.

[†] See the footnote on page 41 for an in-depth explanation of this limit.

[‡] Similar to before, the additional factor of $\frac{1}{2}$ will create a more constructive pattern later on.

But because we are adding twelve triangles, we must also subtract away twelve side lengths. In total, we have

$$p_2 = p_1 + 12 \times 3 \left(\frac{s}{3^2}\right) - 12 \left(\frac{s}{3^2}s\right)$$
$$= p_1 + 2^3 \times 3 \left(\frac{s}{3^2}\right)$$
$$= p_1 + 2^4 \times \frac{3}{2} \left(\frac{s}{3^2}\right).$$

Performing one final iteration ($n = 3$), we add forty-eight more triangles for a total of $3 \times 48 = 144$ new edges, each with length $\frac{1}{3^3}s$. As before, we subtract away the forty-eight superfluous side lengths, making the perimeter

$$p_3 = p_2 + 48 \times 3 \left(\frac{s}{3^3}\right) - 48 \left(\frac{s}{3^3}\right)$$
$$= p_2 + 2^5 \times 3 \left(\frac{s}{3^3}\right)$$
$$= p_2 + 2^6 \times \frac{3}{2} \left(\frac{s}{3^3}\right).$$

In general, it appears that

$$p_{n+1} = p_n + 2^{2(n+1)} \times \frac{3}{2} \left(\frac{s}{3^{n+1}}\right), \qquad (2.30)$$

which is the governing recurrence relationship for the perimeter of the Koch snowflake at the nth iteration.

To unroll the recurrence relationship, first note that from (2.30)

$$p_n = p_{n-1} + 2^{2n} \times \frac{3}{2} \left(\frac{s}{3^n}\right),$$

making

$$p_{n+1} = p_{n-1} + 2^{2n} \times \frac{3}{2} \left(\frac{s}{3^n}\right) + 2^{2(n+1)} \times \frac{3}{2} \left(\frac{s}{3^{n+1}}\right)$$
$$= p_{n-1} + \frac{3}{2}s \left(\frac{2^{2n}}{3^n} + \frac{2^{2(n+1)}}{3^{n+1}}\right).$$

But from (2.30), we know

$$p_{n-1} = p_{n-2} + 2^{2(n-1)} \times \frac{3}{2} \left(\frac{s}{3^{n-1}}\right),$$

§2.5. The Koch Snowflake

implying

$$p_{n+1} = p_{n-2} + 2^{2(n-1)} \times \frac{3}{2} \left(\frac{s}{3^{n-1}}\right) + \frac{3}{2}s\left(\frac{2^{2n}}{3^n} + \frac{2^{2(n+1)}}{3^{n+1}}\right)$$

$$= p_{n-2} + \frac{3}{2}s\left(\frac{2^{2(n-1)}}{3^{n-1}} + \frac{2^{2n}}{3^n} + \frac{2^{2(n+1)}}{3^{n+1}}\right).$$

It is not difficult to see that the overarching pattern is

$$p_{n+1} = p_0 + \frac{3}{2}s\left(\frac{2^2}{3} + \frac{2^4}{3^2} + \frac{2^6}{3^3} + \cdots + \frac{2^{2(n+1)}}{3^{n+1}}\right), \quad (2.31)$$

where $p_0 = 3s$. As before, we can evaluate the sum on the right using the handy formula

$$C + C\omega + C\omega^2 + \cdots + C\omega^n = \frac{C(1-\omega^{n+1})}{1-\omega},$$

which is (2.28). Rewriting the sum in (2.31), we obtain

$$\frac{2^2}{3} + \frac{2^4}{3^2} + \frac{2^6}{3^3} + \cdots + \frac{2^{2(n+1)}}{3^{n+1}}$$

$$= \frac{2^2}{3} + \frac{2^2}{3}\left(\frac{2^2}{3}\right) + \frac{2^2}{3}\left(\frac{2^2}{3}\right)^2 + \cdots + \frac{2^2}{3}\left(\frac{2^2}{3}\right)^n$$

from which we deduce that $C = \omega = \frac{2^2}{3} = \frac{4}{3}$. Hence, by (2.28) we see

$$\frac{2^2}{3} + \frac{2^4}{3^2} + \frac{2^6}{3^3} + \cdots + \frac{2^{2(n+1)}}{3^{n+1}} = \frac{4}{3}\left[\frac{1-\left(\frac{4}{3}\right)^{n+1}}{1-\frac{4}{3}}\right]$$

$$= -4\left[1 - \left(\frac{4}{3}\right)^{n+1}\right].$$

Thus, by (2.31) we have

$$p_{n+1} = p_0 - 6s\left[1 - \left(\frac{4}{3}\right)^{n+1}\right].$$

We are interested in the perimeter p_∞ of the Koch snowflake, which is acquired in the limit as $n \to \infty$. As a result,

$$p_\infty = \lim_{n\to\infty} p_{n+1} = \lim_{n\to\infty}\left(p_0 - 6s\left[1 - \left(\frac{4}{3}\right)^{n+1}\right]\right).$$

But because $\left|\frac{4}{3}\right| > 1$, as $n \to \infty$, $\left(\frac{4}{3}\right)^{n+1} \to \infty$. Accordingly,

$$\lim_{n \to \infty} p_{n+1} = \infty,$$

forcing the conclusion that $p_\infty = \infty$, as well. Consequently, the Koch snowflake, even though its area is finite, has an *infinite* perimeter. This is certainly a challenge to our geometric intuitions, but the math speaks for itself.

Later in §3.4 we will encounter a geometric object that bumps the Koch snowflake's bizarre perimeter and area relation one dimension higher: infinite surface area but finite volume. This is Gabriel's horn.

§2.6 Kelly's Criterion

We close our discussion on limits and differential calculus with the following scenario:

> *You have been approached by a gambler who entices you to play a betting game based on its seemingly unfair odds—your chance of winning each round is 60% while the gambler's is only 40%. The exact details of the game are unimportant to us, but the gambler did emphasize an important point: You are free to play as many rounds as you like, provided you throw up a consistent wager. The question we intend to answer is thus: Given your chances of winning, what percentage of your bankroll should you bet in order to maximize your long-run profits?*

Our analysis begins with a binary betting game (binary because you either win or you lose) in which you intend to bet some fraction ω of your bankroll. Per the gambler's comment, you are free to play this game as many rounds as you like—say n times. The problem asks for the optimal value of ω so that as $n \to \infty$ ("in the long-run") you make money.

§2.6. Kelly's Criterion

Assuming the probability of winning is constant throughout all rounds of the game, we can define some probabilities as follows:

$$p := \text{your probability of winning}$$
$$q := \text{your probability of losing.}$$

As we will discuss in §3.6, the sum of the probabilities of all the outcomes for an event must be exactly one. This is equivalent to saying that the chance of something occurring is always beneath or equal to 100%. Since you either win or lose, it follows from this principle that

$$p + q = 1. \tag{2.32}$$

Let N_0 be the initial amount of money in your bankroll. This way, if you win the first round of betting, your money will increase to the amount $(1+w)N_0$, but if you lose, your money decreases to the amount $(1-w)N_0$. In general, playing the game n times yields a bankroll

$$N = (1+w)^{pn}(1-w)^{qn}N_0.^{\dagger} \tag{2.33}$$

Restating the problem, we wish to find w such that N is maximized. We do this by finding the value of w such that the derivative $\frac{dN}{dw}$ vanishes. Evidently, this requires the derivative $\frac{dN}{dw}$. Though one could apply the definition of the derivative directly on (2.33) to obtain $\frac{dN}{dw}$, this would evolve into an unpleasant algebraic mess. The best way to find this derivative would be to recognize that (2.33) is a product of two functions—namely, $(1+w)^{pn}$ and $N_0(1-w)^{qn}$. This inspires us to search for a general formula equal to the derivative of the product of two differentiable functions $f(x)$ and $g(x)$—that is,

$$\frac{d}{dx}f(x)g(x).$$

[†] The exponents pn and qn are the expected number of times you will win and lose, respectively, over the n iterations of the game.

Such a formula is obtained by manipulating the definition of the derivative on $f(x)g(x)$:

$$\frac{d}{dx}f(x)g(x) = \lim_{\Delta x \to \infty} \frac{f(x+\Delta x)g(x+\Delta x) - f(x)g(x)}{\Delta x}.$$

Adding zero to the numerator in the form of the fancy subtraction $f(x)g(x + \Delta x) - f(x)g(x + \Delta x)$ produces the limit

$$\lim_{\Delta x \to \infty} \frac{f(x+\Delta x)g(x+\Delta x) - f(x)g(x)}{\Delta x}$$
$$+ \lim_{\Delta x \to \infty} \frac{f(x)g(x+\Delta x) - f(x)g(x+\Delta x)}{\Delta x}.$$

Pairing the first and last, and second and third terms in the numerator, and then noting

$$\lim_{x \to \Delta x} g(x + \Delta x) = g(x) \text{ and } \lim_{x \to \Delta x} f(x) = f(x)$$

brings us to the limit

$$f(x)\left[\lim_{\Delta x \to \infty} \frac{g(x+\Delta x) - g(x)}{\Delta x}\right]$$
$$+ g(x)\left[\lim_{\Delta x \to \infty} \frac{f(x+\Delta x) - f(x)}{\Delta x}\right].$$

Reconciling the bracketed limits with the definition of the derivative reveals the handy *derivative product rule*:

$$\frac{d}{dx}f(x)g(x) = f(x)g'(x) + g(x)f'(x). \quad (2.34)$$

Indeed, this will prove to be a serviceable formula both now and later.

We can utilize (2.34) to compute $\frac{dN}{d\omega}$ by assigning $f(\omega) = (1 + \omega)^{pn}$ and $g(\omega) = N_0(1 - \omega)^{qn}$. Applying the product rule and setting the result equal to zero (remember, we want to maximize N) yields

$$\frac{dN}{d\omega} = pn(1+\omega)^{pn-1}(1-\omega)^{qn} - qn(1+\omega)^{pn}(1-\omega)^{qn-1} = 0.$$

§2.6. Kelly's Criterion

Note that the constant N_0 disappears since the expression is equal to zero. Reformulating, we have

$$(1+\omega)q = (1-\omega)p \implies \omega = \frac{p-q}{p+q}.$$

But the sum $p + q = 1$ by (2.32). Therefore,

$$\omega = \frac{p-q}{1} = p - q. \tag{2.35}$$

What we've just derived is a betting strategy called the *Kelly criterion*, named after its architect John Kelly. The interpretation of (2.35) is straightforward: To maximize your long-run profits in a binary betting event, always wager the difference between your chance of winning and chance of losing. To illustrate with the scenario at the top of this section, we had $p = 60\%$ and $q = 40\%$. According to the Kelly criterion, you should bet $60\% - 40\% = 20\%$ of your bankroll to maximize future profits. Notice that with $p = q = 50\%$, (2.35) tells us $\omega = 0\%$. In other words, in any fair binary game you should bet no money whatsoever if your intent is to profit in the long-run. This agrees with our intuitions because in fair games we'd expect to break even, and thus not make (or lose) any money.

Of course, these are games of chance, so there is always a lingering nonzero probability of going broke no matter your odds of winning. Evidently, the only way to guarantee the safety of your money is to never play at all, but this takes the fun out of things.

CHAPTER 3

INTEGRAL CALCULUS

"It is not worth an intelligent man's time to be in the majority. By definition, there are already enough people to do that."

∼ G.H. Hardy

INTEGRAL calculus, like differential calculus, rests upon the notion of a limit. But rather than providing information about the slope of the tangent line along some curve, integral calculus allows us to compute the area underneath such a curve. The significance of this, unfortunately, is hidden away in the geometric interpretation. What we explain here is the unseen relationship between the derivative and something called an *integral*, which is, in some way, an operation that "unfastens" the derivative.

For the quadrillionth time now, we know the derivative of a function $f(x)$ is defined by the limit

$$\frac{d}{dx}f(x) := \lim_{\Delta x \to 0} \frac{f(x + \Delta x) - f(x)}{\Delta x}.$$

In the previous chapter, we interpreted this as being the slope of the tangent line at any point x. Our first goal in this chapter is to describe an operation that undoes the derivative. In other words, we want to find perform some operation on

$$\frac{d}{dx}f(x) = f'(x)$$

so that we get $f(x)$ back. The operator that does this is called the *indefinite integral*.

Symbolically, the indefinite integral of $f(x)$—sometimes called the *antiderivative* of $f(x)$—is notated using the elongated S character

$$\int f(x)dx,$$

where $\int$ is the integral symbol and $f(x)$ is called the *integrand*. The dx will be described in a later paragraph, but it generally tells us with respect to which variable the integral is evaluated (so with dx integrate with respect to x, and with $d\xi$ integrate with respect to ξ).

We introduced the indefinite integral as an operator that reverses a derivative. It is therefore reasonable to suspect that

$$\int f'(x)dx = f(x).$$

This equality, however, is seldom the case. Unfortunately, one can never undo a derivative completely because information about $f(x)$ is lost whenever $f(x)$ contains a constant term (the derivative of a constant vanishes). Consider, for instance, the derivative functions $f'(x) = 2x$ and $g'(x) = 2x$. Just looking at these you may suspect $f(x) = g(x) = x^2$, and so

$$\int f'(x)dx = \int g'(x)dx = x^2.$$

However, it could just as easily be true that $f(x) = x^2 + 1$ while $g(x) = x^2 - 42$. Though both have the same derivative, they differ by a constant. We have thus established the following result:

> If two arbitrary derivative functions $f'(x)$ and $g'(x)$ are equal, it is not necessarily the case that $f(x) = g(x)$.

Therefore, the general formula for antidifferentiation is best expressed as

$$\int f'(x)dx = f(x) + C \tag{3.1}$$

where C is a constant—called the *constant of integration*.

(3.1) provides a nice summary of what the indefinite integral will look like, but it lends no insight for how to actually compute antiderivatives. While there are many ways to do this, our approaches in this book will be one of two. The first involves simply asking the question: Given a function $f'(x)$, what function $f(x)$ has derivative $f'(x)$? This $f(x)$ is the antiderivative of $f'(x)$. This method is called *integration by inspection*.

To illustrate this technique, consider the derivative function

$$f'(x) = x^n.$$

We require a function $f(x)$ such that

$$\frac{d}{dx}f(x) = x^n.$$

Since the derivative decreases the order of a polynomial function by one, we expect the antiderivative of $f(x)$ to be of order $n + 1$. This encourages us to guess the function

$$f(x) = x^{n+1} \implies f'(x) = (n+1)x^n.$$

This derivative is off by a factor of $n+1$. Though this is no problem, we can simply divide the trial function x^{n+1} by $n+1$ to obtain the proper antiderivative:

$$f(x) = \frac{1}{n+1}x^{n+1} \implies f'(x) = x^n.$$

Chapter 3. Integral Calculus

Of course, the same result holds for $f(x) + C$, where C is any constant. Hence,

$$\int x^n \, dx = \frac{1}{n+1} x^{n+1} + C. \tag{3.2}$$

(3.2) will prove to be a useful integration formula later on. Other important ones (which are also derived using inspection) include

$$\int e^x \, dx = e^x + C \quad \text{and} \quad \int \frac{1}{x} \, dx = \log|x| + C,^\dagger$$

and also some trigonometric ones

$$\int \cos(x) \, dx = \sin(x) + C \quad \text{and} \quad \int \sin(x) \, dx = -\cos(x) + C.$$

All of these can be verified by comparing the derivative of the right side to the integrand—you will find that they are equal.

The second technique we use in evaluating integrals is called *integration by substitution* whose purpose is to make a hard integral less so. For instance, the function

$$f'(x) = \frac{x}{x^2 + 1}$$

has no obvious antiderivative. That is, there is no apparent function $f(x)$ such that

$$\frac{d}{dx} f(x) = \frac{x}{x^2 + 1}.$$

Notwithstanding our intimidation by this problem, we perform a type of substitution called a *u-substitution*, where we set the variable $u = x^2 + 1$ and then observe that

$$\frac{du}{dx} = 2x \implies \frac{1}{2} du = x \, dx.$$

Hence the integral

$$\int \frac{x}{x^2 + 1} dx = \int \frac{1}{x^2 + 1} \underbrace{(x \, dx)}_{\frac{1}{2} du} = \int \frac{1}{2u} du.$$

---

†Here, $\log(x) = \ln(x)$ is the *natural logarithm* with base e. This is true of all logarithms in this book. We take the absolute value to ensure the output is real-valued.

This last integral is easily evaluated using the logarithm formula from above:

$$\int \frac{1}{2u} du = \frac{1}{2} \int \frac{1}{u} du = \frac{1}{2} \log|u| + C.$$

And from our substitution $u = x^2 + 1$, it follows that the antiderivative of $f'(x)$ is $\frac{1}{2} \log|x^2 + 1| + C$. Clearly $x^2 + 1 > 0$ for all $x \in \mathbb{R}$. Therefore, we can omit the absolute value sign and conclude with

$$\int \frac{x}{x^2 + 1} dx = \frac{1}{2} \log(x^2 + 1) + C.$$

We will encounter two additional substitution methods in later sections. Until then, we require an understanding of something called a *definite integral*.

The definite integral from a to b of $f(x)$ (notated $\int_a^b f(x) dx$) is defined by something called a *Riemann sum*, which is the limit on the right-hand side below:

$$\int_a^b f(x) dx := \lim_{n \to \infty} \frac{b - a}{n} [f(x_1) + f(x_2) + \cdots + f(x_n)], \quad (3.3)$$

where all $x_1, x_2, \cdots, x_n$ are values in the domain of $f(x)$ such that

$$a < x_1 < x_2 < \cdots < x_n < b.$$

To illustrate (3.3) more geometrically, consider Figure 3.1. Shown are various rectangles of differing height but of identical width Δx. For each value x_j, notice the height of the corresponding rectangle is $f(x_j)$. It follows that if we were to sum

$$\Delta x f(x_1) + \Delta x f(x_2) + \cdots + \Delta x f(x_n)$$
$$= \Delta x [f(x_1) + f(x_2) + \cdots + f(x_n)]$$

we would be adding the areas of each rectangle beneath the curve of the function $f(x)$. In other words, such a sum would approximate the area underneath the function $f(x)$ from $x = a$ to $x = b$. For

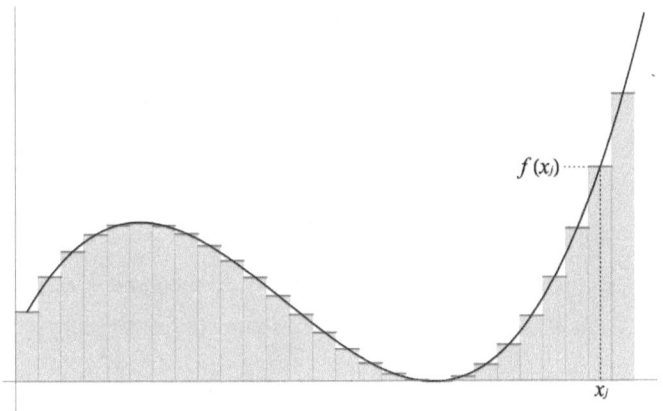

Figure 3.1: Graphical interpretation of a Riemann sum.

n rectangles of equal width situated between the two values $x = a$ and $x = b$, their width Δx is given by the simple formula

$$\Delta x = \frac{b-a}{n}.$$

Hence,

$$\Delta x \big[f(x_1) + f(x_2) + \cdots + f(x_n)\big]$$
$$= \frac{b-a}{n}\big[f(x_1) + f(x_2) + \cdots + f(x_n)\big].$$

Thus as $n \to \infty$, the number of rectangles approaches infinity, but their width Δx becomes infinitesimally small. This brings us back to the dx in the integral sign. The dx is really an infinitesimal version of Δx in the Riemann sum—it is the infinitesimal width of the many rectangles that compose the area beneath a curve. So as Δx (the change in x) becomes smaller and smaller, it approaches dx—an infinitesimal length.

In summary, the definite integral described by

$$\int_a^b f(x)dx$$

outputs a value whose geometric meaning is the area underneath the function $f(x)$ from $x = a$ to $x = b$. From our discussion on indefinite integrals, for a derivative function $f'(x)$ the definite integral over the interval $[a, b]$ is, as you may suspect,

$$\int_a^b f'(x)dx = f(b) - f(a). \tag{3.4}$$

This result is the first of two major pieces in the *Fundamental Theorem of Calculus*. The second part is somewhat a corollary of the first. Setting $b = x$, (3.4) becomes

$$\int_a^x f'(x)dx = f(x) - f(a).$$

Because $f(a)$ is a constant, it follows that

$$\frac{d}{dx}\int_a^x f'(x)dx = \frac{d}{dx}f(x) = f'(x).$$

In other words, the derivative of the integral of a function is the function, implying differentiation and integration are reverse operations.

A final integral worth addressing is a type of definite integral called an *improper integral*. These are integrals whose bounds of integration are actually limits, such as $\pm\infty$. For instance, the integrals

$$\int_a^\infty f(x)dx, \int_{-\infty}^a f(x)dx, \text{ and } \int_{-\infty}^\infty f(x)dx$$

are all improper integrals. The evaluation of such integrals is relatively straightforward. Each bound on the integral can (and should) be thought of as a limit. So the improper integral

$$\int_a^\infty f(x)dx = \lim_{N\to\infty}\int_a^N f(x)dx$$

and can thus be evaluated in terms of N, with the limit taken later on. We will see many examples of improper integrals and this exact procedure in upcoming sections.

§3.1. THE QUADRATIC FORMULA

We now have all the necessary tools to understand the mathematics in this chapter. We'll start with a painless yet powerful application of integral substitution to prove the quadratic formula and work our way up to prove that π is irrational.

§3.1 THE QUADRATIC FORMULA

Perhaps the most famous (infamous?) mathematical expression among all who've ever completed a course in algebra is the *quadratic formula*—the equation governing the roots of the second-order polynomial

$$f(x) := ax^2 + bx + c.$$

The quadratic formula tells us that for

$$x = \frac{-b \pm \sqrt{b^2 - 4ac}}{2a}, \tag{3.5}$$

$f(x) = 0$. This can be verified explicitly, but often a derivation of something is more fulfilling than a simple verification. Thus, in this section we derive (3.5) using the basic principles of integral calculus.

For sake of clarity, we let $x = t$ so that

$$f(x) \to f(t) = at^2 + bt + c.$$

Differentiating with respect to t yields

$$f'(t) = 2at + b.$$

By the fundamental theorem of calculus (part one), the integral of a function's derivative over some interval is that function evaluated between the interval. That is,

$$\int_0^x (2at + b)dt = f(x) - f(0) \implies f(x) = \int_0^x (2at + b)dt + c \tag{3.6}$$

because $f(0) = c$.

Okay, at present all is calm. A magical storm emanates, however, from the clever u-substitution:

$$u = 2at + b \implies du = 2a\,dt.$$

Reconcile these values with (3.6) and the integral becomes

$$f(x) = \int_{b}^{2ax+b} \frac{u}{2a} du + c. \qquad (3.7)$$

Note the change on the bounds of the integral. This is because the integration has changed from $t = 0, t = x$ to $u = 2a(0) + b = b, u = 2ax + b$.

Evaluating (3.7) using (3.2) and the fundamental theorem of calculus, we find

$$f(x) = \frac{(2ax+b)^2}{4a} - \frac{b^2}{4a} + c.$$

We want to find x such that $f(x) = 0$. So, in setting $f(x) = 0$, it follows that

$$0 = \frac{(2ax+b)^2}{4a} - \frac{b^2}{4a} + c \implies x = \frac{-b \pm \sqrt{b^2 - 4ac}}{2a},$$

which is the quadratic formula. While more profound results reside in this chapter, our short analysis here is certainly a stylish means for obtaining the quadratic formula. It is all thanks to areas under curves.

§3.2 Euler's Equation and the Proverbial $e^{i\pi}$

In this section we derive what is arguably the most famous and enticing formula in all mathematics. The equation provides an incredible link between many fundamental corners of mathematics, including trigonometry, exponentiation, basic arithmetic, and complex numbers.

§3.2. Euler's Equation and the Proverbial $e^{i\pi}$

We begin, somewhat arbitrarily, with the function

$$J(\theta) := \cos(\theta) + i\sin(\theta).$$

This formulation was obtained by merely combining a few of the characters we are interested in: trigonometric ratios and the complex unit $i := \sqrt{-1}$. From inception, we see that $J(0) = \cos(0) + i\sin(0) = 1 + i0 = 1$. This will prove useful momentarily. To get a better handle on $J(\theta)$, let's compute the first derivative

$$\frac{d}{d\theta}J(\theta) = -\sin(\theta) + i\cos(\theta).^{\dagger} \qquad (3.8)$$

Recalling the property $i^2 = -1$, we can rewrite (3.8) as

$$\frac{d}{d\theta}J(\theta) = i^2\sin(\theta) + i\cos(\theta)$$
$$= i\underbrace{[i\sin(\theta) + \cos(\theta)]}_{J(\theta)}$$

More explicitly, we have shown

$$\frac{d}{d\theta}J(\theta) = iJ(\theta).$$

This is called a *separable differential equation*, "differential equation" because the expression involves the derivative of $J(\theta)$, and "separable" because it can be algebraically restructured such that different variables are on distinct sides of the equation, as in

$$\frac{dJ(\theta)}{J(\theta)} = i\,d\theta.$$

This looks closely tied to something we integrate, and this is precisely what you do to solve the equation in full:

$$\int \frac{dJ(\theta)}{J(\theta)} = \int i\,d\theta \implies \log(J(\theta)) = i\theta + C, ^{\ddagger}$$

---

†For what we pursue here, the same differentiation and integration rules apply when complex numbers are present.

‡This logarithm is really a *complex logarithm*—a particular function used in the field of complex analysis that allows for negative arguments. For this reason, we have neglected to include the absolute value sign on the logarithm.

where C is the constant of integration and $\log(x)$ is the natural logarithm. Raising both sides to the power of e, we find

$$J(\theta) = e^{i\theta + C}.$$

As noted above, $J(0) = 1$. This initial value allows us to solve for C:

$$1 = e^{i(0)+C} = e^C \implies C = 0$$

and so we have shown $J(\theta) = e^{i\theta}$. Setting this equal to the definition of $J(\theta)$ above constructs the astounding formula

$$e^{i\theta} = \cos(\theta) + i\sin(\theta). \tag{3.9}$$

This incredible formulation is called *Euler's equation* after the great Leonhard Euler. One of the most fascinating corollaries of (3.9) transpires with $\theta = \pi$. Because $\cos(\pi) = -1$ and $\sin(\pi) = 0$, it follows that

$$e^{i\pi} + 1 = 0. \tag{3.10}$$

Take a moment to embrace the profundity here.

Commonly called *Euler's identity*, (3.10) relates the most noble constants in mathematics: e—the constant whose exponential form is its own derivative, π—the ratio of a circle's circumference to its diameter, i—the complex unit equal to $\sqrt{-1}$, one—the multiplicative identity ($a \times 1 = a$), and zero—the additive identity ($a + 0 = a$). This expression is the connection between all of these constants. Many concur that (3.10) is the most elegant equality in mathematics.

With something as startling as (3.10), we would be remiss not to mention another fascinating consequence of (3.9). Instead of setting $\theta = \pi$, consider $\theta = \frac{\pi}{2}$. Under this setting, $\cos\left(\frac{\pi}{2}\right) = 0$ and $\sin\left(\frac{\pi}{2}\right) = 1$. Hence,

$$e^{i\pi/2} = i.$$

Raising both sides to the power of i yields

$$i^i = e^{i^2 \pi/2} = e^{-\pi/2}. \tag{3.11}$$

In more quantitative terms, $i^i = \sqrt{-1}^{\sqrt{-1}}$ is a *real* number, given approximately as

$$i^i = e^{-\pi/2} \approx 0.20788.$$

With all these unbelievable results, a word of negligence is in order. Notice how the trigonometric functions sine and cosine in (3.9) are periodic after the interval $[0, 2\pi)$. Consequently, the equality $e^{i\theta} = \cos(\theta) + i\sin(\theta)$ is periodic on the same interval. This means that not only is $e^{i\pi} = -1$, but also that $e^{i(3\pi)} = -1$, $e^{i(5\pi)} = -1$, and so forth. In general, for any odd $k \in \mathbb{Z}$,

$$e^{ik\pi} + 1 = 0.$$

So while (3.10) is true for $k = 1$, it is also true for infinitely many more k.

A more pressing concern from the periodic nature of $e^{i\theta}$ is (3.11). It is true that for all $k \in \mathbb{Z}$,

$$\cos\left(\frac{\pi}{2}(4k+1)\right) = 0 \text{ and } \sin\left(\frac{\pi}{2}(4k+1)\right) = 1.$$

Therefore,

$$e^{i(4k+1)\pi/2} = i \implies i^i = e^{-(4k+1)\pi/2}.$$

In other words, i^i is not a well-defined value because it depends on the choice of k.[†] Regardless, the choice $k = 0$ seems to be preferred upon entering i^i on most calculators. We will discuss both (3.9) and (3.10) again in §4.2, and also address the issue of choosing k in i^i and similar expressions. Until then, we have more integrals to evaluate.

§3.3 THE PARABOLIC NATURE OF FREE FALL

[†]Evidently, i^i remains a real number regardless of k. This is a surprise in itself.

Calculus was conceived within the minds of Isaac Newton and Gottfried Wilhelm Leibniz, and born at the stroke of their pens. Unwittingly, humanity stumbled upon the most versatile field of mathematics to date. For Newton, his calculus was necessary to formalize the axioms of motion and provide the first mathematically consistent theory of gravity—two of the most significant advancements in all of physics. As a result of this origin, it should come with little surprise that both calculus and physics are fundamentally intertwined—one is always helping the other.

We find one of the most beautiful (albeit elementary) applications of calculus to be the prediction of parabolic flight for free falling objects. Moreover, Newton's calculus and his laws of motion argue that all objects, in the absence of air and friction, fall at the same rate under gravitational influence irrespective of mass and size. It is the purpose of this section to prove that all objects fall at the same rate under gravity as well as the parabolic nature of free fall.

We start with Newton's three laws of motion, which are perhaps the most important physical statements ever devised. The first goes like this:

> *An inertial body remains inertial unless acted on by an outside force.*

Here *inertial* is taken to mean either at rest (i.e. the body is not moving) or at constant velocity. In either case, the body is not accelerating, so its velocity does not change. Now instead of nodding our heads at this law in mild negligence, let's take a moment to truly grasp the counterintuitive nature of this claim. We are familiar with moving things always coming to a halt, such as a book sliding across a table or a skateboard rolling freely on a flat road. Where have you ever seen an object, with no motor or any type of propulsion, move indefinitely as this law postulates? Your answer is the same

§3.3. The Parabolic Nature of Free Fall

as mine—never. Of course, unbeknownst to Newton's compatriots, we are familiar with the reason these objects slow down—friction. More precisely, it is the frictional *force* that slowly, but surly, retards the speed at which an object travels.

The notion of force brings us to Newton's second law of motion:

> *An object's acceleration is proportional to the force applied and inversely to its mass.*

Symbolically, Newton's second law states $a := \frac{F}{m}$, where m is the mass of some object, F is the force being applied to that object, and a is the object's resulting acceleration. In other words, an object's *acceleration* is defined as the force applied to an object divided by its mass (in a moment we will encounter a separate, more intuitive definition of acceleration). Inherently, we are more familiar with the reformulation

$$F = ma. \tag{3.12}$$

This, as you know, is one of the most famous equations ever written. We will uncover a small mystery of the universe by employing this elegant equation shortly. But first, there is one more law of motion to cover.

In its most popular syntax, Newton's third law postulates:

> *For every action there is an equal and opposite reaction.*

In other words, if a force is applied to an object by you, then the object supplies the same force back onto you. The quintessential mechanics of a rifle are relevant here. The bullet is propelled out of the barrel because an exothermic chemical reaction has applied a very large force on the bullet. In return, the bullet supplies an equal and opposite force to the casing in which the chemical reaction took place, which forces the entire gun into the shoulder of the shooter—a phenomenon commonly referred to as the "kick" of the rifle. Such a kick is the equal and opposite reaction. We are now ready to explore a simple yet profound consequence of these axioms.[†]

On top of his calculus and his three laws of motion, Newton showed that planets, which can be approximated as perfect spheres, behave like point masses in space and thus exhibit a gravitational force on an object with mass m according to the equation

$$F = -\frac{GM_p m}{r^2}, \text{[‡]} \tag{3.13}$$

where M_p is the mass of the planet (for Earth, $M_p = M_\oplus \approx 5.98 \times 10^{24}$ kg),[§] r is the distance from the center of mass m to the center of the planet (with Earth as M_p, we will assume the mass m is on the surface, making $r = R_\oplus$ the radius of the Earth—about 6.37×10^6 m), and $G \approx 6.67 \times 10^{-11}$ m^{3}kg^{-1}s^{-2} is the *gravitational constant*. The negative sign in (3.13) signifies that the gravitational force is an attractive one—a phenomenon to which we are highly accustomed.

Utilizing (3.12) in the context of Newton's law of gravity formulates an astounding consequence. Since both (3.12) and (3.13) are representations for F, we can set them equal to each other to find

$$ma = -\frac{GM_p m}{r^2} \implies a = -\frac{GM_p}{r^2}. \tag{3.14}$$

Notice how the mass m cancels out. Though trivial to note mathematically, this algebraic cancellation remains a deep mystery in physics. That said, we can infer the cancellation implies that under

---

[†]We state this with caution, as these formulations are subject to change when velocities are very near the speed of light $c = 299,792,458$ ms^{-1}. Here the work of Einstein and his theory of relativity need be employed to accurately detail physical motion. We provide a brief introduction to these ideas in §5.6.

[‡]*Historic Note*: (3.13) is the precise equation NASA scientists used to chart the path of the Saturn V rocket during the Apollo program. In light of its fame, a proof that (3.13) is applicable to spherical planets is provided in §A.3.

[§]The symbol $\oplus$ is the astronomical symbol for *Terra* (a.k.a. the planet Earth).

§3.3. The Parabolic Nature of Free Fall

pure gravitational effects, *all objects accelerate at the same rate, regardless of their mass.* Naturally, this is counterintuitive because our intuitions have been molded with the pervasive influence of drag and other forces for which mass matters a great deal. Nonetheless, (3.14) is an intrinsic law of the cosmos.

Employing (3.14) we can compute the rate at which objects accelerate near the surface of Earth. Inputting the gravitational constant G and the mass and radius of Earth provided above, we compute

$$a = g \approx -9.8 \text{ ms}^{-2}, \tag{3.15}$$

where g is the acceleration due to gravity at the Earth's surface. Because the mass of the object is independent of this quantity, all objects on the surface of the Earth, absent all other forces like drag, will accelerate down towards the surface at 9.8 ms^{-2}. This is a monumental realization—an object's gravitational acceleration is solely dependent on the mass it interacts with, not the mass of the object itself. It was this idea, along with many other insights, that laid the groundwork for Albert Einstein when he superseded Newton's law of gravity with his general theory of relativity—the topic of §5.6.

We now transition to see how this insight—that all objects accelerate at the same rate regardless of mass and size—implies free falling objects trace out parabolic paths. We begin with a definition: *Acceleration is the rate of change of velocity.* If v is velocity and a acceleration, then this definition is equivalent to writing

$$a := \frac{dv}{dt}.$$

This equality is a separable differential equation, like that encountered in the last section. Separating the variables by moving the dt over to the other side and integrating, we obtain

$$\int dv = \int a\, dt \implies v = at + C_1,$$

where C_1 is the constant of integration. If we set $t = 0$, it follows that $v = C_1$. Physically, this implies that C_1 is the object's initial velocity, which we denote v_0. This is a reasonable substitution because, intuitively, dropping an object from rest will take longer to reach a high speed than one moving very fast to begin with. If we make one final adjustment to the equation and label v as $v(t)$—parametrizing the velocity as a function of time t—our expression for velocity becomes

$$v(t) = at + v_0. \tag{3.16}$$

This is called the *kinematic equation for velocity*, and is familiar to nearly any student who has taken a physics course.

The definition of velocity is the rate of change of position—that is,

$$v(t) := \frac{ds}{dt},$$

where s denotes the position of the falling object. Substituting (3.16) into this definition we obtain the separable differential equation

$$\frac{ds}{dt} = at + v_0.$$

Separating variables and integrating, we find

$$\int ds = \int (at + v_0)dt \implies s(t) = \frac{1}{2}at^2 + v_0 t + C_2.$$

Similar to before, we intuit that the integration constant C_2 should represent the initial position s_0 of the object. Consequently,

$$s(t) = \frac{1}{2}at^2 + v_0 t + s_0, \tag{3.17}$$

which is the *kinematic equation for position*. (3.17) describes the position as a function of time for any object with initial velocity v_0 that is subject to a constant acceleration a.

To prove parabolic flight, we now imagine a cannon positioned on horizontal ground that aims up some angle β from the horizontal, as in Figure 3.2. Suppose the cannon fires a projectile at an initial

§3.3. The Parabolic Nature of Free Fall

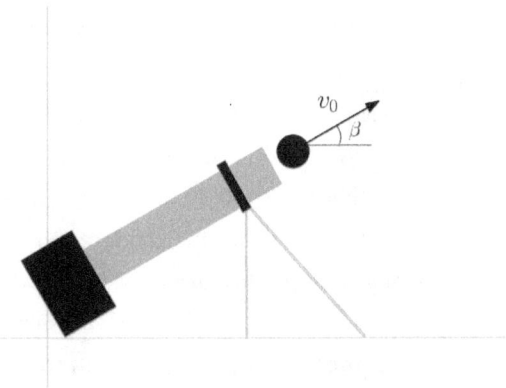

Figure 3.2: A cannon firing a projectile with velocity v_0 at an angle β to the horizontal.

velocity v_0 up and to the right at the angle β. This velocity is directional, so we can divide it up into its horizontal and vertical components. Using Figure 3.2, the projectile has a horizontal component of velocity

$$v_x = v_0 \cos(\beta) \tag{3.18}$$

and a vertical component of velocity

$$v_y = v_0 \sin(\beta).$$

Divvying up the velocity in this way may seem abnormal at first. We'll find, however, that there is great power in doing this. As can be confirmed by merely throwing a ball upward in a car, the motion of an object in one spatial direction is independent of the object's motion in a separate, perpendicular direction. So whether you're moving at 80 mph on the freeway or standing still on Fenway Park, a baseball tossed vertically upward will travel up and down back into your hand, irrespective of your horizontal speed.[†] This fact is a consequence of Newton's first law. In the car you are in an inertial

---

[†] Again, assuming no acceleration.

(non-accelerating) frame, and so the ball behaves exactly as it would had you not been moving.

Going back to the cannon and with the new insight that we can treat the projectile's vertical and horizontal motion independently, we start by contemplating what the acceleration is in each direction. A consequence of (3.15) is that the acceleration due to gravity always pulls orthogonally to the surface of Earth, which, in the cannon problem, is parallel to the y-axis. Thus, the acceleration of the projectile in the y-direction is just $a_y = g$. In the x-direction, we acknowledge that, unlike the y-direction, no external forces act on the projectile, so by Newton's second law there is no horizontal acceleration—that is, $a_x = 0$.

Substituting this information into (3.17), the x component of position become

$$s_x(t) = v_x t + s_x.$$

The coordinate system delineating the cannonball's motion is completely arbitrary. And since the initial position of the cannonball is an artifact of the coordinate system, it is also arbitrary. We can therefore choose to orient the coordinate axes such that the ball's initial position is zero—that is, so $s_x = s_y = 0$. Combining this with the expression for v_x in (3.18), we have that

$$s_x(t) = v_0 \cos(\beta) t + 0 \implies t = \frac{s_x(t)}{v_0 \cos(\beta)}. \qquad (3.19)$$

Doing the same for the y-direction except with with $a_y = g$, we obtain

$$s_y(t) = \frac{1}{2} g t^2 + v_0 \sin(\beta) t.$$

But we know from (3.19) what the parameter t is. Using this, we conclude that the cannonball's position in space is governed by the quadratic polynomial

$$s_y(t) = \frac{g \sec^2(\beta)}{2v_0^2} s_x^2(t) + \tan(\beta) s_x(t).^\dagger \qquad (3.20)$$

This is a second-order polynomial in $s_x(t)$, and it depicts a parabola on the x, y-plane. This then proves that all free falling objects follow parabolic trajectories under pure gravitational influence.

Naturally, if the cannonball's velocity is very great it may be sent into orbit around the Earth, which is obviously not a parabolic trajectory. This illustrates that the preceding analysis breaks down for those circumstances wherein g changes throughout the flight. That said, these equations are enticing nonetheless, and are certainly applicable to the ball with which many play catch.

§3.4 GABRIEL'S HORN

In just three sections integral calculus has proven to be quite a versatile discipline. In this section we bolster such versatility by outlining an application of definite integrals beyond the computation of areas to volumes of three-dimensional objects.

Our method for computing volumes relies heavily on *solids of revolution*, which are three-dimensional objects constructed by rotating a two-dimensional graph around a specified line. We illustrate the idea with the linear function

$$f(x) := 2x \text{ on the interval } [0, 4],$$

which is plotted in Figure 3.3.

Making use of our imaginations, visualize a small rectangle like that shaded in Figure 3.3 rotating about the y-axis a full revolution.

---

†The attentive reader notes (3.20) resembles an upward opening parabola—certainly not the trajectory to which free falling objects succumb. Remember that $g < 0$ as noted in (3.15). This means the first term in (3.20) is negative, thus producing the downward facing parabola we expect.

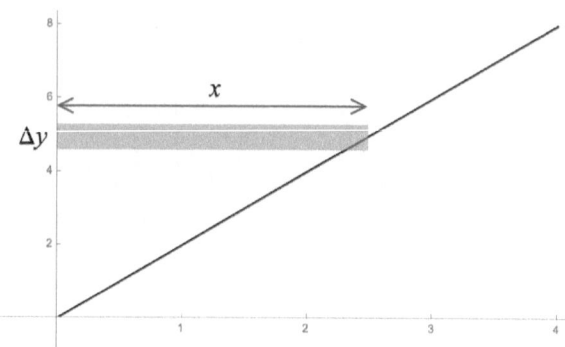

Figure 3.3: The graph of $f(x) = 2x$ on the interval $[0, 4]$.

The resulting shape would be a solid disk of radius x and thickness Δy. Imagine now rotating the entire function $f(x) = 2x$ about the y-axis so that it pivots on its lower end at the origin. The resulting shape is the cone in Figure 3.4. The swept out rectangle, now a disk of thickness Δy, is analogous to a rectangle in a Riemann sum. Evidently, here the resulting sum is a volume and not an area.

To compute the volume V of the cone above, we can use the general integral formula

$$V = \int v(y).$$

Here $v(y)$ is a function in y for the volume of the individual disks that makeup Figure 3.4. We previously detailed that the swept out disk has radius x and height Δy. Therefore, the volume of the disk is $\pi x^2 \Delta y$. And by the function $f(x) = y = 2x$, we have that $x = \frac{y}{2}$ making our volume function

$$v(y) = \frac{\pi y^2}{4} \Delta y.$$

In the infinitesimal limit $\Delta \to 0$, $\Delta y \to dy$ making the volume integral

$$V = \int \frac{\pi y^2}{4} dy.$$

§3.4. Gabriel's Horn

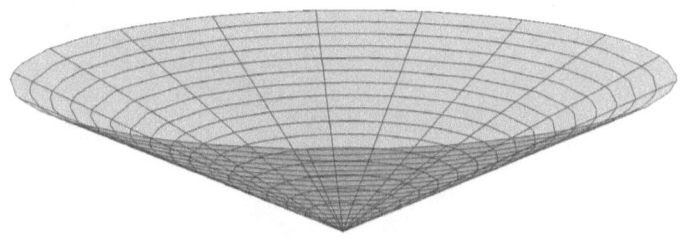

Figure 3.4: The revolution of $f(x) = 2x$ on $[0, 4]$ about the y-axis.

Finally, because $x \in [0, 4]$ and $y = 2x$, it follows that $y \in [0, 8]$. Hence, the total volume of the swept out cone is the definite integral

$$V = \int_0^8 \frac{\pi y^2}{4} dy = \frac{\pi}{12} y^3 \Big|_0^8 = \frac{128\pi}{3}.$$

Whether interesting or not, this example illustrates the general process for proving two astounding features of a solid of revolution called *Gabriel's horn*. This object is named not after a mathematician but rather a more divine being—the angel Archangel Gabriel. This is because, similar to the Koch snowflake, the horn comprises some otherworldly features—its volume is finite (so its interior could be filled with water) but its surface area is infinite (so its exterior could never be fully painted). Envisioning such an object is difficult, but Figure 3.5 lends some insight into the shape of Gabriel's horn.

The horn is derived via a rotation of the function

$$G(x) := \frac{1}{x} \text{ on the interval } [1, \infty)$$

about the x-axis. To prove its finite volume, we will first derive the volume function $v(x)$ for individual disks along the x-axis. The radius of each disk, a distance x from the origin, will be $y = \frac{1}{x}$, and

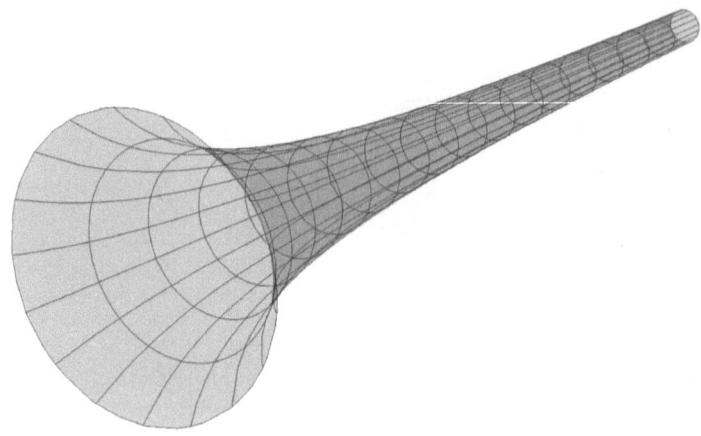

Figure 3.5: Gabriel's Horn.

have a thickness Δx. It follows that

$$v(x) = \frac{\pi}{x^2}\Delta x,$$

which, in the infinitesimal limit, is

$$v(x) = \frac{\pi}{x^2}dx.$$

Hence, the volume of Gabriel's horn is

$$V = \int v(x) = \int_1^\infty \frac{\pi}{x^2}dx = -\frac{\pi}{x}\bigg|_1^\infty = \pi,$$

a finite value.

The surface area is a similar calculation, but instead we seek a function $a(x)$ for the surface area of each disk that makes up the horn. More specifically, by surface area of each disk we mean the edge surface area (the portion a wheel rolls on) and not the round top and bottom portions (the rim of the wheel). For a disk of radius y and thickness Δx, simple geometry tells us the surface area of this outer edge is $2\pi y \Delta x$. Noting $y = \frac{1}{x}$, it follows that

$$a(x) = \frac{2\pi}{x}\Delta x,$$

which, in the infinitesimal limit, is

$$a(x) = \frac{2\pi}{x} dx.$$

Consequently, the total surface area of Gabriel's horn is

$$A = \int a(x) = \int_1^\infty \frac{2\pi}{x} dx = 2\pi \log |x| \Big|_1^\infty = \infty,$$

which, obviously, is an infinite value.

Therefore, Gabriel's horn—a geometric object derived through the rotation of $\frac{1}{x}$ on $[1, \infty)$ about the x-axis—is a solid of finite area (equal to π) and infinite surface area. Like the Koch snowflake in §2.5, Gabriel's horn is yet another peculiar consequence of infinity and a challenge to our geometric intuitions.

§3.5 COINS AND CYCLOIDS

Whether through the notion of infinity (e.g. Gabriel's horn) or highly abstract ideas (e.g. seven-dimensional spheres), intuition can be challenged in numerous ways. In this section, we hope to test your intuition by performing a simple experiment with which nearly everyone is familiar—rolling a few coins on a table.

Envision two identical coins, call them C_1 and C_2, laid down one right next to the other, as in Figure 3.6. If you roll C_2 about the center of C_1 (so it rolls along the outer circumferences of C_1), through how many radians must C_2 traverse to get all the way around C_1, back to its starting position? Almost immediately, you may fire back with the following argument:

- Let the radius of both coins be R.

- The distance the outer coin C_2 must travel is the circumference of the center coin C_1, which is $2\pi R$.

- And because C_2 also has radius R and needs to roll through a distance $2\pi R$—the length of its circumference—it will rotate

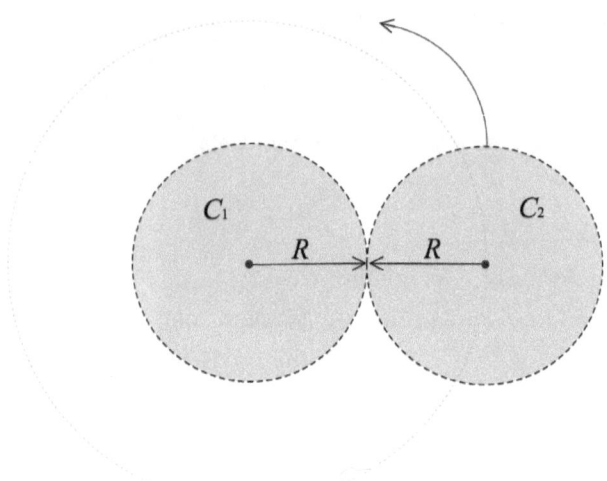

Figure 3.6: Two identical coins C_1 and C_2 of radius R, where C_2 rolls around C_1.

through one revolution, which is 2π radians, to get all the way back around.

Yet upon trying the experiment, as if we were physicists, practice does not agree with theory.

The error in our reasoning lies in our interpretation of the motion of the outer coin C_2. Specifically, we presumed we could "unravel" the circumference of the inner coin C_1, and simply roll the outer coin C_2 along a straight track of length $2\pi R$. Of course, then, C_2, being a disk of radius R, must subtend 2π radians in order to traverse the distance. But this is only works because the *center* of the coin C_2 travels a distance $2\pi R$. If the center travels a longer distance, then the coin will clearly rotate through more revolutions. This is the insight needed to deduce the correct answer to the problem—focus not on the edge of the coin, but on its center. Approach the problem this way and the true answer follows

§3.5. COINS AND CYCLOIDS

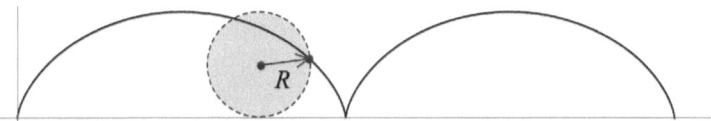

Figure 3.7: A cycloid curve constructed from a disk of radius R.

immediately.

To see why, consider again Figure 3.6. Notice the center of C_2 is a distance $2R$ from the center of C_1. Hence, rolling C_2 around C_1 causes the center of C_2 to trace out a circle of radius $2R$, meaning that in a full revolution, C_2 will travel a distance $2\pi(2R) = 4\pi R$. Thus, C_2 actually rolls through 4π revolutions to get around the center coin—twice the answer we previously computed, and also the result obtained by experimentation.

To some this answer may have been obvious from the start, either from experience or great intuition. But let's now analyze a related, more difficult problem. Consider one of the coins from above, but now with a small spot of ink somewhere on its circumference. Imagine rolling the coin next to a piece of paper so that the ink continually rubs off on the paper. The arc traced out would resemble a shape called a *cycloid*, like that in Figure 3.7.

Cycloids are fascinating creatures. Not only is an inverted cycloid the solution to the famous Brachistochrone problem, which asks to construct the curve between two points such that a ball rolling down the curve travels between the end points in the least possible time, but cycloids also hold a somewhat counterintuitive (at least unexpected) length of arc. It is the purpose of this section to deduce the length of a cycloid arc, in relation to the radius R of the circle that made it.

Similar to the coin problem, let's try to guess the solution for the arclength of a cycloid before we actually compute it. We are

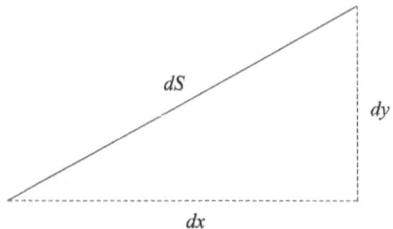

Figure 3.8: An infinitesimal length of arc dS.

interested in the region where the circle goes through one full revolution. In Figure 3.7, this corresponds to the length of one of the two mounds. Because the curve is traced out by a circle at a point on its circumference, rotating the circle a full revolution will correspond to a curve whose arclength is $2\pi R$. While this is true when the circle is stationary, we will soon see that the arc is a bit longer due to the fact that the circle is not only rotating, but is also moving forward.

To deduce the exact length of arc, we essentially need to add up infinitesimal arclengths of the cycloid. In other words, if S is the arclength of the cycloid, then, somewhat heuristically, we want to evaluate the integral

$$S = \int dS, \tag{3.21}$$

where dS is an infinitesimal length of arc (remember, the integral is a sum of infinitesimal lengths). We can deduce dS by considering what an infinitesimal length dS would look like on the cycloid.

Zooming in on any continuous curved line will make it appear linear (this idea is emblematic of calculus). Hence, if dx is a small distance in the x-direction and dy a small distance in the y-direction, then up close dS will resemble the line segment in Figure 3.8 with catheti dx and dy. It follows from the Pythagorean theorem that

$$dS^2 = dx^2 + dy^2 \implies dS = \sqrt{dx^2 + dy^2}, \tag{3.22}$$

§3.5. COINS AND CYCLOIDS

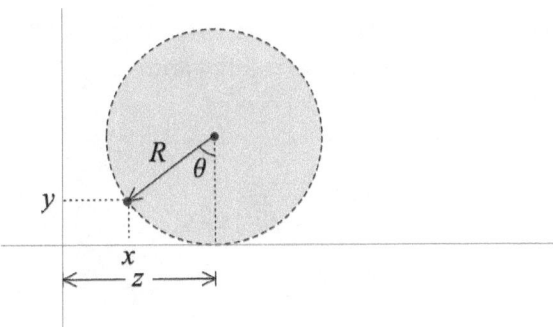

Figure 3.9: A graphic for deducing a parametrization in θ for the cycloid.

where we have taken the positive solution because length is always a positive measure. Of course, (3.22) is only of use to us if we know some information about dx and dy, which, in turn, requires an understanding of x and y. To obtain this information, we will deduce the general equation of a cycloid.

To do this, consider Figure 3.9, which represents the beginning of a cycloid construction by rolling a disk of radius R along the x-axis. Here, $z = R\theta$ is the distance traveled by the center of the disk from the origin (as with any cycloid, we assume the generating disk rolls without slipping). After contemplating the graphic for a bit, it becomes clear that $x = z - R\sin(\theta)$ while $y = R - R\cos(\theta)$. That is,

$$\begin{cases} x(\theta) = R\theta - R\sin(\theta) \\ y(\theta) = R - R\cos(\theta). \end{cases} \quad (3.23)$$

Indeed, graphing these *parametric equations*—a set of equations in terms of a common variable, in this case θ—yields the cycloidal arc we are after.

Recall our motivation for wanting (3.23): We need some idea of what the differential lengths dx and dy look like. Now it may seem somewhat problematic that (3.23) is in terms of θ, but some algebraic magic on (3.22) will readily resolve this worry:

$$dS = \sqrt{dx^2 + dy^2} \left(\frac{d\theta}{d\theta}\right)$$

$$= \sqrt{\frac{dx^2 + dy^2}{d\theta^2}}\, d\theta$$

$$= \sqrt{\left(\frac{dx}{d\theta}\right)^2 + \left(\frac{dy}{d\theta}\right)^2}\, d\theta.$$

Hence, the arclength for one mound of the cycloid (a complete revolution of the generating circle) is, by (3.21), the integral

$$S = \int dS = \int_0^{2\pi} \sqrt{\left(\frac{dx}{d\theta}\right)^2 + \left(\frac{dy}{d\theta}\right)^2}\, d\theta.^\dagger$$

But the ratios $\frac{dx}{d\theta}$ and $\frac{dy}{d\theta}$ are the derivatives of the functions in (3.23), which are straightforward to evaluate:

$$\begin{cases} x'(\theta) = \frac{dx}{d\theta} = R - R\cos(\theta) \\ y'(\theta) = \frac{dy}{d\theta} = R\sin(\theta). \end{cases}$$

Therefore,

$$\left(\frac{dx}{d\theta}\right)^2 + \left(\frac{dy}{d\theta}\right)^2 = (R - R\cos(\theta))^2 + (R\sin(\theta))^2$$

$$= R^2 - 2R^2\cos(\theta) + \underbrace{R^2\cos^2(\theta) + R^2\sin^2(\theta)}_{R^2}$$

$$= 2R^2 - 2R^2\cos(\theta)$$

$$= 2R^2\left(1 - \cos(\theta)\right),$$

and so the arclength S is just the integral

$$S = \int_0^{2\pi} \sqrt{2R^2\left(1 - \cos(\theta)\right)}\, d\theta = \sqrt{2}R \int_0^{2\pi} \sqrt{1 - \cos(\theta)}\, d\theta.$$

---

†We evaluate the integral from $\theta = 0$ to $\theta = 2\pi$ because this is the range of θ that corresponds to a single mound in the cycloid arc.

§3.5. Coins and Cycloids

This integral is begging for a substitution. Our choice is a u-substitution on $\cos(\theta)$ so that

$$u = \cos(\theta) \implies du = -\sin(\theta)d\theta.$$

Now because $1 - \cos^2(\theta) = \sin^2(\theta)$, it follows that $\sin^2(\theta) = 1 - u^2$ and so $\sin(\theta) = \pm\sqrt{1-u^2}$. This makes

$$d\theta = \mp\frac{du}{\sqrt{1-u^2}}.^\dagger$$

Now we must be careful from here on out. The sine function is positive for all $\theta \in (0, \pi)$, negative for all $\theta \in (\pi, 2\pi)$, and zero for $\theta = 0, \pi, 2\pi$. Therefore, to encapsulate the entire integration range with our substitution (from $\theta = 0$ to $\theta = 2\pi$), we must break the integral up into the two regions $0 < \theta < \pi$ (this is where $\sin(x)$ is positive) and $\pi < \theta < 2\pi$ (this is where $\sin(x)$ is negative).

With this substitution, the arclength becomes

$$S = \sqrt{2}R \int_{\theta=0}^{\theta=\pi} \overbrace{-\frac{\sqrt{1-u}}{\sqrt{1-u^2}}}^{\sin(\theta)>0} du + \sqrt{2}R \int_{\theta=\pi}^{\theta=2\pi} \overbrace{\frac{\sqrt{1-u}}{\sqrt{1-u^2}}}^{\sin(\theta)<0} du.$$

Thankfully these have identical integrands (except for the sign) so their evaluations are identical. Notice that the fraction

$$\frac{\sqrt{1-u}}{\sqrt{1-u^2}} = \sqrt{\frac{(1-u)}{(1-u)(1+u)}} = \sqrt{\frac{1}{1+u}}.$$

Consequently,

$$S = \sqrt{2}R \int_{\theta=0}^{\theta=\pi} -\sqrt{\frac{1}{1+u}}\, du + \sqrt{2}R \int_{\theta=\pi}^{\theta=2\pi} \sqrt{\frac{1}{1+u}}\, du.$$

These integrals are far simpler than when we started. They evaluate nicely to yield

$$S = \sqrt{2}R\left[-2\sqrt{1+u}\right]_{\theta=0}^{\theta=\pi} + \sqrt{2}R\left[2\sqrt{1+u}\right]_{\theta=\pi}^{\theta=2\pi}.$$

$\dagger$ The symbol $\mp$ is the minus/plus symbol. It results when the plus/minus symbol $\pm$ is multiplied by a negative number.

Reconciling this solution with the substitution $u = \cos(\theta)$, we obtain

$$S = \sqrt{2}R \left[-2\sqrt{1+\cos(\theta)}\right]_0^\pi + \sqrt{2}R \left[2\sqrt{1+\cos(\theta)}\right]_\pi^{2\pi},$$

which equates to

$$S = \sqrt{2}R \left[2\sqrt{2}\right] + \sqrt{2}R \left[2\sqrt{2}\right]$$
$$= 4R + 4R$$
$$= 8R.$$

In other words, the arclength of one mound of a cycloid is $8R$, where R is the radius of the cycloid's generating circle. Because $2\pi \approx 6.28$, it is clear that $8R > 2\pi R$. This confirms our hypothesis that the arclength of a cycloid is larger than the circumference of its generating circle.

What is often difficult to grasp is the simplicity in the solution for the arclength. Even though the cycloid is generated by a circle, which makes it seem like the arc of the cycloid should be intimately related to π, it is not. Rather, the arclength S is a simple, rational number (assuming R is rational) given by $S = 8R$. Quite remarkable, to say the least.

§3.6 THE GAUSSIAN FUNCTION

In this section we pursue a very omnipresent function in the field of *statistics*—that which studies randomness and chance. A notion acutely embedded within statistics is *probability*—the likelihood of something occurring, such as flipping a coin and it landing on heads. More formally, for an event in which only a discrete number of outcomes are possible, the probability of a favorable outcome (e.g. heads over tails) is the ratio

$$\text{probability} := \frac{\text{number of favorable outcomes}}{\text{number of possible outcomes}}. \tag{3.24}$$

§3.6. The Gaussian Function

For the contrived coin example there are two outcomes: tails and heads. We want heads (the favorable outcome), thus by (3.24) the probability of obtaining heads is $\frac{1}{2}$—in agreement with common knowledge.

One of the reasons probability is such an important concept is its definition requires it to always be within zero and one. This is because there must always be an amount of favorable outcomes equal to or less than the total number of possible outcomes. Though it may not be so obvious, this realization closely associates statistics with calculus. The relation emanates from the fact that calculus is very helpful in formulating functions whose definite integral represents the probability that some event will occur. It should come as no surprise, then, that deeply embedded within statistics are the underlying principles of integral calculus. But before venturing any deeper into this relationship, it will be useful to first define a few statistical terms.

Statisticians define a *probability distribution* as a function that encodes the probability for an event in some statistical experiment. To illustrate, consider again the experiment of flipping a fair coin. Here, only two outcomes are feasible: heads or tails, each with the same 50% chance of occurring. A suitable probability distribution for this event would then be the piecewise function

$$T(x) := \begin{cases} \text{heads:} & \frac{1}{2} \\ \text{tails:} & \frac{1}{2}, \end{cases}$$

where x is either heads or tails. Notice that in adding the probability of each possible outcome, the net sum is one. And this fact is true in general: For any event with a discrete number of outcomes, the sum of the probabilities for all outcomes is exactly one. To put this more formally, let $P(x)$ be an arbitrary *discrete probability distribution* defined as the piecewise function

$$P(x) := \begin{cases} \text{outcome 1:} & \alpha_1 \\ \text{outcome 2:} & \alpha_2 \\ \quad \vdots & \\ \text{outcome } n: & \alpha_n, \end{cases}$$

where x could be any of the n possible outcomes. Here, each α_j denotes the probability that the jth outcome is brought to fruition. Hence, because α_j is a probability we require $0 \leq \alpha_j \leq 1$. Not only this, but the addition rule stated above implies the sum

$$\alpha_1 + \alpha_2 + \alpha_3 + \cdots + \alpha_n = 1. \tag{3.25}$$

The statistical meaning of (3.25) is simple: For any statistical event with a discrete number of feasible outcomes, one of these outcomes must occur. As intuitively obvious this may appear, it is crucial in order to grasp the motivation behind this section.

We now transition from the more mundane discrete statistical experiment, to the continuous one. Often times probability distributions need to be formulated such that they are fluid and continuous, meaning there are an infinite number of possible outcomes confined within a particular range. Imagine throwing a dart at a ruler one foot in length and being interested in the value to which the dart becomes affixed. Assuming this ruler is infinitely precise and that the tip of each dart is infinitesimally minute, the ruler essentially becomes the number line on the interval $I = [0, 1]$ (in feet). Since there are infinitely many real numbers in this interval I, there is an infinite number of possible outcomes for the experiment.

This leads us to an important point: With continuous distributions, it is aimless to pose questions concerning discrete probability distributions such as, "What is the probability of throwing a dart

§3.6. THE GAUSSIAN FUNCTION

and it sticking, say, 0.342 feet?" By the discrete probability definition in (3.24), we can compute this probability. There is one favorable outcome (sticking 0.342) out of the infinity of possible outcomes. Therefore, $p = \frac{1}{\infty} \to 0$. Of course, this result is not unique to 0.342—the probability of a discrete outcome in a continuous distribution is invariably zero.

In order to obtain nonzero probabilities in continuous distributions, we must inquire about probabilities regarding continuous regions. To illustrate, consider again the dart game described above. Instead of asking what is the probability of hitting a particular value, we ask, "What is the probability of throwing a dart and it sticking in the interval, say, $[0.2, 0.4]$?" Here this question has a nonzero answer, related not to individual outcomes, but the range of outcomes in which the dart-thrower is interested. The probability of this occurring is

$$p := \frac{\text{favorable region}}{\text{total region}} = \frac{0.4 - 0.2}{1 - 0} = 0.2.$$

The probability defined above is that for a *continuous probability distribution* for which there are many varieties, the most popular being the *uniform* and *normal distributions*. The uniform distribution is that associated with the dart throwing game above, and is simply a probability function that remains a constant value over the interval of possible outcomes (for the dart game, this was the interval $[0, 1]$). The uniform distribution, though, is not nearly as interesting as the normal distribution, which is shown in Figure 3.10 and whose probability function is defined as

$$\phi(x) := \frac{1}{\sigma\sqrt{2\pi}} \exp\left(-\frac{(x-\mu)^2}{2\sigma^2}\right). \tag{3.26}$$

Here the function $\exp(x)$ (the exponential of x) carries the same meaning as e^x, where $e = 2.718281\cdots$ is Euler's number. The symbol μ denotes the mean, or average, of the set of data being

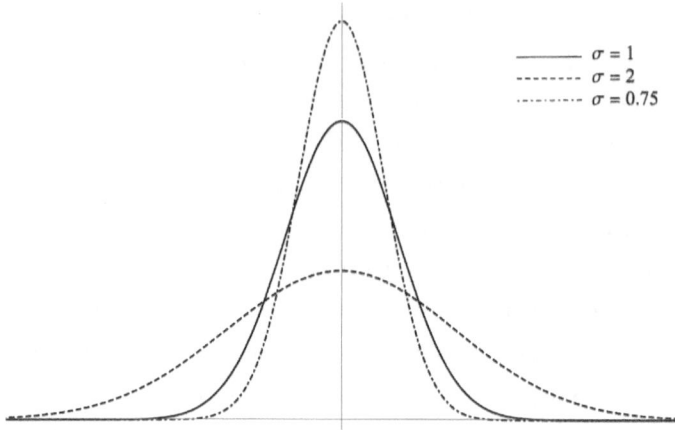

Figure 3.10: The normal distribution with $\mu = 0$ and varying σ.

described by the distribution and the symbol σ denotes the standard deviation of the data, which is a metric for the data's spread about the mean μ.

For the special case with $\mu = 0$ and $\sigma = 1$ (the undashed line in Figure 3.10) statisticians call the corresponding distribution the *standard normal distribution*. As seen by the two other dashed curves, altering σ modifies the shape of the distribution. Though not indicated, variations in the mean μ shift the graph so that the maximum point on the curve is centered above μ.

What is so fascinating about all the possible curves that can be generated from (3.26) (and also any other continuous distribution) is that each embodies the probabilistic axiom of always summing to unity. Here the "sum" is the integral over the region on which the probability distribution is defined. Since the normal distribution is defined for all $x \in (-\infty, \infty)$, the total area beneath the curve must, by this axiom, be exactly one. In other words,

$$\int_{-\infty}^{\infty} \phi(x)dx = 1 \qquad (3.27)$$

§3.6. THE GAUSSIAN FUNCTION

for any choice of μ and σ. This should come as a surprise given the awkward shape of (3.26), but we will shortly see that this is indeed the case. To do so, we begin with an introduction to *Gaussian functions*.

A Gaussian function, named after the acclaimed seventeenth-eighteenth century mathematician Karl Friedrich Gauss, is one that houses an e^{-x^2} term. At a loss of creativity, mathematicians call the *Gaussian integral* that over the simplest Gaussian function—namely,

$$I := \int_{-\infty}^{\infty} e^{-x^2} dx. \tag{3.28}$$

It turns out this integral does not have a closed form, meaning it cannot be expressed in terms of elementary functions[†] unless integrated over the entire number line as in (3.28). Thankfully, (3.27) is also being integrated over this interval, so we should be able to assign a bona fide value to it. Rather than integrate $\phi(x)$ directly, however, it will make the algebra a bit easier if we instead compute the Gaussian integral I (a monster by itself), then apply what we learn to treat (3.27).

To evaluate (3.28), first recognize that the function e^{-x^2} is an *even* function, meaning it is symmetric about the y-axis (see Figure 3.10 to get an idea). This implies that

$$I = 2 \int_0^{\infty} e^{-x^2} dx, \tag{3.29}$$

[†]Elementary functions are those composed of a *finite* number of arithmetic operators ($+, -, \times, \div$), trigonometric functions, exponentials/logarithms, and roots of polynomial equations. We show in §A.4 that

$$\int_{-\infty}^{\infty} e^{-x^2} dx = \lim_{x \to \infty} \left(x - \frac{x^3}{3 \cdot 1!} + \frac{x^5}{5 \cdot 2!} - \frac{x^7}{7 \cdot 3!} + \cdots \right),$$

where the $\cdots$ implies the addition and subtraction goes on forever. This suggests (though not definitively) that (3.28) is not an elementary function because there is an infinite use of addition and subtraction in its evaluation.

because the area below the function e^{-x^2} to the left of the y-axis is equal to the area below e^{-x^2} to the right of the y-axis. Furthermore, notice that the variable x is a *dummy variable*, which means it can be replaced by a separate variable and (3.29) will still convey the same value. In other words,

$$I = 2 \int_0^\infty e^{-x^2} dx = 2 \int_0^\infty e^{-y^2} dy$$

for some other variable y. From this, it follows that

$$I^2 = \left(2 \int_0^\infty e^{-x^2} dx\right) \left(2 \int_0^\infty e^{-y^2} dy\right).$$

Relative to the integral with respect to y, the quantity given by

$$2 \int_0^\infty e^{-x^2} dx \qquad (3.30)$$

is just a constant, so we can bring it inside the integral

$$2 \int_0^\infty e^{-y^2} dy$$

like you do with a numerical constant. Thus,

$$I^2 = 2 \int_0^\infty \left(2 \int_0^\infty e^{-x^2} dx\right) e^{-y^2} dy.$$

Similarly, the outside integral (the integral with respect to y) is a constant relative to the inside integral. This means that we can bring everything inside (3.30) to formulate the double integral

$$I^2 = 4 \int_0^\infty \int_0^\infty e^{-x^2} e^{-y^2} dy dx = 4 \int_0^\infty \int_0^\infty e^{-(x^2+y^2)} dy dx. \qquad (3.31)$$

To many, (3.31) looks worse to evaluate than just the single integral I. But we can greatly simplify the computation through the substitution

$$x = yt \implies dx = y dt.$$

§3.6. The Gaussian Function

This change of variables transforms (3.31) to

$$I^2 = 4 \int_0^\infty \int_0^\infty y e^{-y^2(1+t^2)} dy\, dt. \tag{3.32}$$

Because we are first integrating with respect to y (the dy is the innermost differential), the quantity $1 + t^2$ is treated as constant. Hence, the inside integral becomes

$$\int_0^\infty y e^{-ay^2} dy,$$

with $a = 1 + t^2$. Making a u-substitution with

$$u = ay^2 \implies \frac{1}{2a} du = y\, dy,$$

we have

$$\int_0^\infty y e^{-ay^2} dy \to \int_0^\infty \frac{1}{2a} e^{-u} du = \frac{1}{2a} = \frac{1}{2(1+t^2)}$$

where in the last step we unfastened the substitution $a = 1+t^2$. We have now evaluated the inside integral of (3.32). Placing this result back into (3.32) we obtain

$$I^2 = 4 \int_0^\infty \frac{1}{2(1+t^2)} dt = 2 \int_0^\infty \frac{dt}{1+t^2}.$$

To evaluate what remains, we employ a particular u-substitution called a *trigonometric substitution*. For this, we let

$$t = \tan(\theta) \implies dt = \sec^2(\theta) d\theta, \tag{3.33}$$

from which it follows that

$$I^2 = 2 \int_0^\infty \frac{dt}{1+t^2} = 2 \int_{t=0}^{t=\infty} \frac{\sec^2(\theta)}{1+\tan^2(\theta)} d\theta = 2 \int_{t=0}^{t=\infty} d\theta$$

because $1+\tan^2(\theta) = \sec^2(\theta)$. Notice from (3.33) that $\theta = \arctan(t)$. Therefore, the integration yields

$$I^2 = 2\left[\arctan(\infty) - \arctan(0)\right] = \pi \implies I = \pm\sqrt{\pi}.$$

Because the function $e^{-x^2} > 0$ for all $x \in (-\infty, \infty)$, the integral of e^{-x^2} over any region is positive. Consequently, we omit the negative solution and happily conclude that

$$I = \int_{-\infty}^{\infty} e^{-x^2} dx = \sqrt{\pi}. \tag{3.34}$$

Though this integral required quite the treatment to solve, the outcome is simple and elegant. It is truly astonishing that π is the sole value in this result, even though a circle is suitably out of sight.

To finalize this section, recall the motivation behind it: Proving that the total area under every normal distribution is one. That is,

$$\int_{-\infty}^{\infty} \phi(x) dx = 1, \tag{3.35}$$

where $\phi(x)$ is defined in (3.26). Under the substitution

$$v^2 = \frac{(x-\mu)^2}{2\sigma^2} \implies x = v\sigma\sqrt{2} + \mu,$$

it follows that $dx = \sigma\sqrt{2} dv$. Combining this with the fact that $\exp(x) = e^x$, (3.35) becomes

$$\frac{1}{\sigma\sqrt{2\pi}} \int_{-\infty}^{\infty} \sigma\sqrt{2} e^{-v^2} dv = \frac{1}{\sqrt{\pi}} \int_{-\infty}^{\infty} e^{-v^2} dv. \tag{3.36}$$

But we know from (3.34) that

$$\int_{-\infty}^{\infty} e^{-v^2} dv = \sqrt{\pi},$$

so it follows from (3.36) that indeed

$$\int_{-\infty}^{\infty} \phi(x) dx = 1.$$

Hence, the area under all possible normal distributions is exactly one regardless of the mean μ and standard deviation σ. Normal distributions, and Gaussian functions in general, are a fascinating bunch indeed. For those interested in studying quantum mechanics, these functions will become your best friend.

§3.7 THE GAMMA FUNCTION

In §2.3 we introduced the *factorial function* $n!$ defined by the product

$$n! := n \times (n-1) \times (n-2) \times \cdots \times 2 \times 1. \qquad (3.37)$$

Despite its simplicity, the factorial function has many applications, most notably in describing the number of ways to permute a quantity of distinct objects. For example, given three letters $A, B,$ and C, there are precisely six ways to distinctly list them:

$$ABC, ACB, BAC, BCA, CAB, \text{ and } CBA.$$

Simple arithmetic tells us that $3! = 6$ is the number of separate ways to list these letters. And this is true in general: Given n discrete objects, there are $n!$ distinct ways to arrange them.

From this standpoint, being asked to compute, say,

$$\left(\frac{1}{2}\right)! \qquad (3.38)$$

is synonymous with asking how many ways $\frac{1}{2}$ objects can be arranged. But what does it mean to have $\frac{1}{2}$ objects? Sure it is possible to have half of an object (such as half of a cake), but not a number of objects whose object-total is fractional (by itself, half of a cake remains a whole object). Clearly, then, arranging $\frac{1}{2}$ objects is an absurd notion. So in order to prevent this and other, similarly irrational queries from being posed, the domain of (3.37) is restricted for $n \in \mathbb{N}$, ensuring $n!$ is operated only on a discrete number of objects.

To most, such a restriction is plausible given the relationship between the factorial function and object arrangement. To others, however, it still appears unnecessarily exclusive. After all, you may be familiar with the fact that $0! = 1$, even though this neither follows from (3.37) nor is it even allowed by the domain $n \in \mathbb{N}$ (remember $0 \in \mathbb{N}_0$ but $0 \notin \mathbb{N}$). Thus, with this apparent discrepancy in the domain of the factorial function, it is feasible to at least try in computing (3.38) with a calculator or some other computer system. In doing so, as if everything we've said about the domain of $n!$ is wrong, a value is generated:

$$\left(\frac{1}{2}\right)! = 0.8862269255\cdots. \qquad (3.39)$$

This value is an interesting creature. While it does not disprove the argument on the impossibility of arranging $\frac{1}{2}$ objects, it does imply the factorial function has a broader application beyond the notion of object arrangement, and, in turn, a larger domain than previously established. It is the purpose of this section to find out exactly what the number in (3.39) is and, in general, to compute all factorials of the form

$$\left(\frac{m}{2}\right)!,$$

where $m \in \mathbb{N}$ and is odd (that is, factorials with $n = \frac{1}{2}, \frac{3}{2}, \frac{5}{2}, \cdots$). We do this by employing the very famous gamma function, whose intimate relationship with the factorial function is thoroughly unforeseen.

By definition, the *gamma function* is the integral

$$\Gamma(n) := \int_0^\infty e^{-x} x^{n-1} dx, \qquad (3.40)$$

where Γ is the capital Greek letter gamma. For now, we assume a domain in (3.40) of $n \in \mathbb{N}$. As we will see momentarily, the factorial function (3.37) is actually embedded within (3.40) itself. The relationship is most readily established by evaluating $\Gamma(n)$ using *integration by parts*.[†]

[†]For those not familiar, integration by parts is a powerful twist on u-substitution.

§3.7. THE GAMMA FUNCTION

Let
$$u = x^{n-1} \implies du = (n-1)x^{n-2}dx$$
$$dv = e^{-x}dx \implies v = -e^{-x}.^\dagger$$

Then
$$\Gamma(n) = \int_0^\infty u\,dv = uv\Big|_0^\infty - \int_0^\infty v\,du.$$

Back substituting, we obtain
$$\Gamma(n) = \underbrace{-x^{n-1}e^{-x}\Big|_0^\infty}_{\text{this term vanishes}} + (n-1)\int_0^\infty e^{-x}x^{n-2}dx$$
$$= (n-1)\int_0^\infty e^{-x}x^{n-2}dx.$$

But by the definition of $\Gamma(n)$ in (3.40),
$$\int_0^\infty e^{-x}x^{n-2} = \Gamma(n-1).$$

Here's a brief introduction to the idea. Given two functions of x, say $u(x)$ and $v(x)$ (hereafter referred to as u and v, respectively), the derivative product rule (see §2.6) tells us that
$$(uv)' = uv' + vu'.$$
By the fundamental theorem of calculus (part one),
$$uv\Big|_a^b = \int_a^b (uv)'\,dx \implies uv\Big|_a^b = \int_a^b uv'\,dx + \int_a^b vu'\,dx.$$
Rearranging, we have that
$$\int_a^b uv'\,dx = uv\Big|_a^b - \int_a^b vu'\,dx. \tag{3.41}$$
In this text we use the more compact notation $du = u'\,dx$ and $dv = v'\,dx$, making (3.41)
$$\int_a^b u\,dv = uv\Big|_a^b - \int_a^b v\,du. \tag{3.42}$$
This is the general formula for integration by parts on the functions $u(x)$ and $v(x)$. Into (3.42) we can substitute any u and dv we like and evaluate the left-hand side integral by the formula on the right-hand side.

†Here the result $v = -e^{-x}$ is derived from evaluating $\int e^{-x}dx$ and ignoring the constant of integration (the constant vanishes because this expression is later treated as a definite integral). You are encouraged to verify this integral.

It follows that
$$\Gamma(n) = (n-1)\Gamma(n-1).$$

Performing integration by parts on $\Gamma(n-1)$, it follows from an identical process that

$$\Gamma(n-1) = (n-2)\Gamma(n-2) \implies \Gamma(n) = (n-1)(n-2)\Gamma(n-2).$$

In general, we obtain

$$\Gamma(n) = (n-1) \times (n-2) \times \cdots \times 3 \times 2 \times \Gamma(1),$$

where
$$\Gamma(1) = \int_0^\infty e^{-x} x^0 dx = \int_0^\infty e^{-x} dx = 1.$$

Consequently, for $n \in \mathbb{N}$

$$\Gamma(n) = (n-1)!. \tag{3.43}$$

This is an important result. Not only does (3.43) demonstrate the relationship between $\Gamma(n)$ and $n!$, but it allows us to redefine the factorial function in terms of $\Gamma(n)$. The consequences of this are enormous since the domain of $\Gamma(n)$ is certainly defined for many other values beyond $\mathbb{N}$. Hence, $\Gamma(n)$ is the key to extending the domain of the factorial function and figuring out the value in (3.39). From here on out, we use the definition

$$n! := \Gamma(n+1). \tag{3.44}$$

To see the power in this new definition, we'll first extend the domain of the factorial function to $\mathbb{N}_0$. By (3.44), $0! = \Gamma(1)$. But this is just the integral

$$\int_0^\infty e^{-x} dx,$$

which we computed above to be one. Therefore, $0! = 1$—in agreement with what we are commonly told in school.

§3.7. THE GAMMA FUNCTION

To extend the domain further, we need only employ (3.44). Consider
$$\Gamma\left(1+\frac{1}{2}\right) = \left(\frac{1}{2}\right)!.$$

By the definition of $\Gamma(n)$,
$$\left(\frac{1}{2}\right)! = \int_0^\infty e^{-x} x^{\frac{1}{2}}\, dx. \tag{3.45}$$

Though this looks quite tricky, a sneaky substitution will make this otherwise daunting integral elementary. Let
$$x = t^2 \implies dx = 2t\, dt.$$

Then
$$\left(\frac{1}{2}\right)! = 2\int_0^\infty e^{-t^2} t^2\, dt.$$

Things are looking worse than (3.45), but through one more round of substitutions (in the form of integration by parts) we will have our answer. Let
$$u = 2t \implies du = 2\, dt$$
$$dv = te^{-t^2}\, dt \implies v = -\frac{1}{2}e^{-t^2},$$

where the result $v = -\frac{1}{2}e^{-t^2}$ is derived from evaluating
$$\int te^{-t^2}\, dt = -\frac{1}{2}e^{-t^2}$$

and ignoring the constant of integration (again, the constant vanishes because this expression is later treated as a definite integral). From these substitutions, we have
$$\left(\frac{1}{2}\right)! = \int_0^\infty u\, dv = uv\Big|_0^\infty - \int_0^\infty v\, du.$$

Back substituting, we obtain
$$\left(\frac{1}{2}\right)! = \underbrace{-te^{-t^2}\Big|_0^\infty}_{\text{this term vanishes}} + \int_0^\infty e^{-t^2}\, dt.$$

Therefore,
$$\left(\frac{1}{2}\right)! = \int_0^\infty e^{-t^2} dt. \qquad (3.46)$$

From the previous section on the Gaussian function, we know the integral

$$\int_{-\infty}^\infty e^{-t^2} dt = \sqrt{\pi} \implies \int_0^\infty e^{-t^2} dt = \frac{\sqrt{\pi}}{2}$$

where the implication follows from e^{-t^2} being an even function. Using (3.46), we have shown

$$\left(\frac{1}{2}\right)! = \frac{\sqrt{\pi}}{2}. \qquad (3.47)$$

In (3.39) we mention the computer output for $\left(\frac{1}{2}\right)!$ is $0.88622\cdots$. Equating $\frac{\sqrt{\pi}}{2}$, we obtain an identical decimal expansion.

Of course, $\left(\frac{1}{2}\right)!$ is a special case in a class of similar factorials—namely, those of the form $\left(\frac{m}{2}\right)!$ where $m \in \mathbb{N}$ and is odd. To evaluate these, we make use of (3.40) and (3.44) to establish that

$$\left(\frac{m}{2}\right)! = \Gamma\left(1 + \frac{m}{2}\right) = \int_0^\infty e^{-x} x^{\frac{m}{2}} dx.$$

This integral is evaluated in a manner identical to that in deriving (3.47). First, we make the substitution

$$x = t^2 \implies dx = 2t\,dt,$$

prompting the result

$$\left(\frac{m}{2}\right)! = 2\int_0^\infty e^{-t^2} t^{m+1} dt.$$

We then employ integration by parts with the substitutions

$$u = 2t^m \implies du = 2mt^{m-1} dt$$
$$dv = te^{-t^2} dt \implies v = -\frac{1}{2} e^{-t^2}.$$

§3.7. The Gamma Function

Back substituting formulates the integral

$$\left(\frac{m}{2}\right)! = \int_0^\infty u\,dv = uv\Big|_0^\infty - \int_0^\infty v\,du = m\int_0^\infty e^{-t^2}t^{m-1}\,dt.$$

Performing another round of integration by parts with

$$u = mt^{m-2} \implies du = m(m-2)t^{m-3}\,dt$$
$$dv = te^{-t^2}\,dt \implies v = -\frac{1}{2}e^{-t^2},$$

we obtain the result

$$\left(\frac{m}{2}\right)! = m \times \frac{(m-2)}{2}\int_0^\infty e^{-t^2}t^{m-3}\,dt.$$

In general, numerous rounds of integration by parts will yield

$$\left(\frac{m}{2}\right)! = m \times \frac{(m-2)}{2} \times \frac{(m-4)}{2} \times \cdots \times \frac{3}{2} \times \frac{1}{2} \times \underbrace{\int_0^\infty e^{-t^2}\,dt}_{\text{this is (3.46)}}.$$

From (3.47) we know $\left(\frac{1}{2}\right)! = \frac{\sqrt{\pi}}{2}$, which also equals (3.46). Therefore,

$$\left(\frac{m}{2}\right)! = m \times \frac{(m-2)}{2} \times \frac{(m-4)}{2} \times \cdots \times \frac{3}{2} \times \frac{1}{2} \times \frac{\sqrt{\pi}}{2}$$
$$= \sqrt{\pi}\left(\frac{m}{2} \times \frac{(m-2)}{2} \times \frac{(m-4)}{2} \times \cdots \times \frac{3}{2} \times \frac{1}{2}\right).$$

Notice there are $\frac{m+1}{2}$ factors in this product (excluding the factor $\sqrt{\pi}$). Therefore, put more succinctly,

$$\left(\frac{m}{2}\right)! = \sqrt{\pi}\left(\frac{m \times (m-2) \times \cdots \times 3 \times 1}{2^{(m+1)/2}}\right). \qquad (3.48)$$

We have successfully extended the domain of the factorial function to, quite literally, infinitely many more values. That said, there is no reason to stop here. In utilizing the definition of the gamma function it is possible to further this extension for nearly all $n \in \mathbb{Q}$. Naturally, the integrals that originate will not be trivial, but the point is valid nonetheless.

We'll close with a final note on notation. The factorial notation $n!$ can be modified to account for the product $n \times (n-2) \times \cdots \times 3 \times 1$ like that in (3.48). Mathematicians define

$$n!! := n \times (n-2) \times \cdots \times 3 \times 1$$

for when n is odd and

$$n!! := n \times (n-2) \times \cdots \times 2 \times 1$$

for when n is even. These are called *double factorials*, and often help shorten lengthy products. Using the double factorial formula for odd n, we can rewrite (3.48) as

$$\left(\frac{m}{2}\right)! = \sqrt{\pi} \left(\frac{m!!}{2^{(m+1)/2}}\right),$$

which makes the overall product a bit more wieldy.

§3.8 The Catenary

So far, many of our calculus adventures have started from basic principles and developed into something reasonably unforeseen. This is certainly the agenda here, where we will utilize our understanding of integration to construct an unexpected function whose arc is nevertheless familiar.

Our exact problem is as follows:

> *To what shape (i.e. function) do hanging chains, power lines, ropes, etc. succumb under pure gravitational influence?*

Your intuition may guide you to believe that such an arc is a parabola (Figure 3.11 surely looks like one). While this is close, we will soon see that the precise arc is mildly more complex.

To properly address the question, we require an extension of calculus into what is called the *calculus of variations*. This is concerned

§3.8. The Catenary

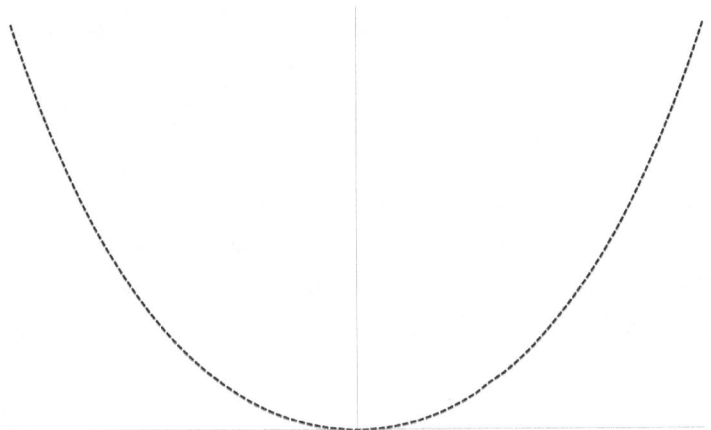

Figure 3.11: The curve to which a hanging chain succumbs under gravity.

not with maximizing or minimizing a particular function (this we know how to do with the derivative) but instead with finding the function that maximizes or minimizes a particular constraint.[†]

To start our analysis, we require a short dialogue concerning a new set of trigonometric functions—the *hyperbolic trigonometric functions*. Similar to how the familiar trigonometric functions (like sine and cosine) are related to the unit circle $x^2 + y^2 = 1$ through the expression $\sin^2(x) + \cos^2(x) = 1$, the hyperbolic trigonometric functions are related to the *unit hyperbola* $x^2 - y^2 = 1$ via

$$\cosh^2(x) - \sinh^2(x) = 1, \qquad (3.49)$$

where $\cosh(x)$ (pronounced "cawsh") is the hyperbolic cosine func-

---

[†]Though the exact details are beyond this book, we would be remiss not to mention the most famous problem associated with the calculus of variations—the *Brachistochrone problem* (this we mentioned back in §3.5). The Brachistochrone problem asks to construct the curve between two points along which an object will move under gravity in less time than any other curve. The answer, in fact, is not a straight line, but rather an inverted cycloid.

tion and sinh(x) (pronounced "sinch") is the hyperbolic sine function. And similar to the regular trigonometric functions, the ratio

$$\frac{\sinh(x)}{\cosh(x)} = \tanh(x)$$

is the hyperbolic tangent function (pronounced "tanch").[†]

Unlike the regular trigonometric functions, the hyperbolic functions have precise definitions based not on geometry and ratios, but on mathematical formulae. To give you three, the hyperbolic cosine is defined as

$$\cosh(x) := \frac{e^x + e^{-x}}{2}, \tag{3.50}$$

the hyperbolic sine as

$$\sinh(x) := \frac{e^x - e^{-x}}{2}, \tag{3.51}$$

and the hyperbolic tangent as

$$\tanh(x) := \frac{\sinh(x)}{\cosh(x)} = \frac{e^x - e^{-x}}{e^x + e^{-x}}.$$

Similar to how sine and cosine can simplify integrals via substitutions, certain characteristics of the hyperbolic functions (the first being (3.49)) allow for a new brand of trigonometric substitutions. Two additional properties include

$$\frac{d}{dx}\sinh(x) = \cosh(x) \text{ and } \frac{d}{dx}\cosh(x) = \sinh(x),$$

which are easily verified by the definitions in (3.50) and (3.51). We will use these shortly to simplify an otherwise difficult integral.

To derive the equation for the arc of a hanging chain (whose ends, we assume, are at an equal height relative to the ground), we require Newton's second law of motion from §3.3:

$$F = ma.$$

[†]In fact, all the common trigonometric functions correspond directly to the hyperbolic ones. For instance, $\frac{1}{\sinh(x)} = \operatorname{csch}(x)$ is the hyperbolic cosecant similar to how $\frac{1}{\sin(x)} = \csc(x)$ is the cosecant.

§3.8. The Catenary

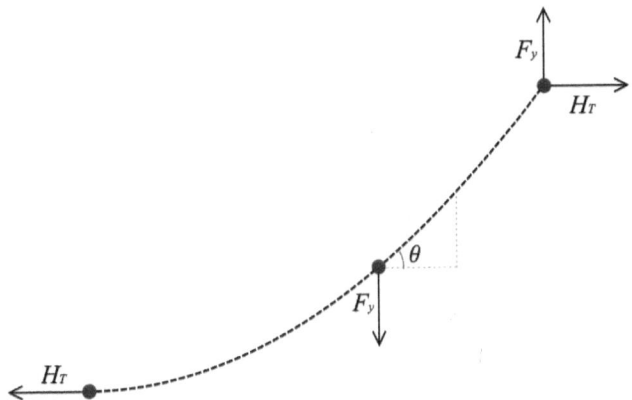

Figure 3.12: A section of a hanging chain with vertical and horizontal forces F_y and H_T, respectively.

Here F is an outside force, m is the mass of some body, and a is that body's acceleration as a result of the force F. On the diagram in Figure 3.12, we have labeled all the forces pulling on some section of the chain. The force F_y in the center is the net force due to gravity acting on the entire segment shown, while the bottom-left force H_T is the horizontal tension throughout the chain. Because hanging chains do not move when at rest (absent other forces caused by wind, etc.), all forces must cancel out so that the chain does not accelerate. In other words, the chain is in a static equilibrium, like the rope with which two equally sized athletes play tug-of-war. Consequently, there must be forces opposing the weight of the chain as well as the horizontal tension—these are provided by the end points of the rope (the top right arrows in Figure 3.12).

In summary, we've determined that at any point along the segment of the chain, the vertical and horizontal forces at that point must cancel. To illustrate more mathematically, consider the center point shown in Figure 3.12. The downward force here is F_y while

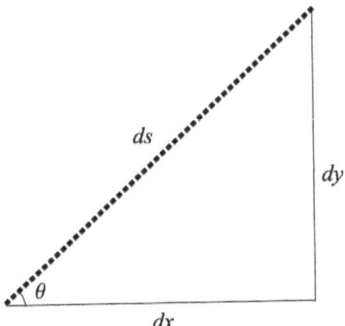

Figure 3.13: An infinitesimal length ds of the hanging chain.

the horizontal force is H_T. Since the chain is in equilibrium, at this position there must also be vertical and horizontal forces with magnitude F_y and H_T, respectively, to cancel the downward and horizontal forces. In Figure 3.12, notice that at the marked point the opposing forces make an angle θ to the tangent of the arc. Curiously, this relates the forces to the geometry of the arc by

$$\tan(\theta) = \frac{F_y}{H_T}.$$

To make this more useful, suppose now that the chain has a total length l and weight w. If $s(x)$ is some amount of arclength of the chain, then $F_y = \frac{w}{l} s(x)$ and so

$$H_T \tan(\theta) = \frac{w}{l} s(x). \tag{3.52}$$

The small differential length ds (shown in Figure 3.13) is at the same point where the tangent makes an angle θ with the horizontal. It is easy to see that

$$\tan(\theta) = \frac{dy}{dx}.$$

Therefore, (3.52) becomes

$$H_T \frac{dy}{dx} = \frac{w}{l} s(x).$$

§3.8. THE CATENARY

Differentiating both sides with respect to x generates

$$H_T \frac{d^2y}{dx^2} = \frac{w}{l} s'(x)$$
$$= \frac{w}{l} \frac{ds}{dx}.$$

Observe from Figure 3.13 that

$$ds = \sqrt{(dx^2) + (dy)^2}$$
$$= \sqrt{1 + \left(\frac{dy}{dx}\right)^2} \, dx$$

from which it follows that

$$H_T \frac{d^2y}{dx^2} = \frac{w}{l} \sqrt{1 + \left(\frac{dy}{dx}\right)^2}.$$

This equation looks very difficult to deal with, but with a small substitution the equation becomes exceptionally cleaner. Let

$$z = \frac{dy}{dx} \implies \frac{dz}{dx} = \frac{d^2y}{dx^2}$$

so that

$$H_T \frac{dz}{dx} = \frac{w}{l} \sqrt{1 + z^2}.$$

This is now just a familiar, single-order differential equation. Isolating the variables, we have

$$\int \frac{dz}{\sqrt{1 + z^2}} = \int \frac{w}{lH_T} dx. \tag{3.53}$$

Solving the right-hand side of (3.53) is no big deal. The left-hand side, however, lacks the same integration luxury. Now is an adequate time to introduce our new-founded hyperbolic functions. We will make the substitution

$$z = \sinh(\nu) \implies dz = \cosh(\nu) d\nu.^\dagger$$

---

†As you can probably guess, a substitution of this form is called a *hyperbolic substitution*.

Thus,
$$\int \frac{dz}{\sqrt{1+z^2}} = \int \frac{\cosh(\nu)}{\sqrt{\cosh^2(\nu)}} d\nu = \int d\nu$$

where in the second step we used the hyperbolic identity $\cosh^2(\nu) = 1 + \sinh^2(\nu)$. Evaluating the differential equation in completion, we find

$$\nu = \frac{w}{lH_T}x + C_1 \implies \operatorname{arcsinh}(z) = \frac{w}{lH_T}x + C_1,$$

where C_1 is the constant of integration. Though never stated explicitly, Figure 3.11 places our hanging chain directly over the origin such that at $x = 0$, $\frac{dy}{dx} = 0$.[†] Hence, by the relationship $\frac{dy}{dx} = z$, it follows that at $x = 0$, $z = 0$. Substituting this into the expression above allows us to determine C_1:

$$\operatorname{arcsinh}(0) = \frac{w}{lH_T}(0) + C_1 \implies 0 = C_1.$$

Consequently, we have deduced that

$$z = \sinh\left(\frac{w}{lH_T}x\right).$$

To finish up, recall that $z = \frac{dy}{dx}$. It follows that

$$\frac{dy}{dx} = \sinh\left(\frac{w}{lH_T}x\right) \implies \int dy = \int \sinh\left(\frac{w}{lH_T}x\right) dx.$$

Using the derivative property $\frac{d}{dx}\cosh(x) = \sinh(x)$, we have

$$\int \sinh(x) dx = \cosh(x) + C_2,$$

where C_2 is yet another constant of integration. Thus,

$$\int \sinh\left(\frac{w}{lH_T}x\right) dx = \frac{lH_T}{w}\cosh\left(\frac{w}{lH_T}x\right) + C_2,$$

[†]This is a consequence of our coordinate system in relation to the position of the hanging chain. Evidently, the coordinate system used to describe the chain is completely arbitrary. Some (like ours) just have easier algebra than others.

from which it follows that

$$y = \frac{lH_T}{w} \cosh\left(\frac{w}{lH_T}x\right) + C_2.$$

Because we are solely concerned with the shape the curve traces out, we can dismiss the constant C_2 (this constant specifies the curve's position on the y-axis, which, for our purposes, is immaterial). Setting $C_2 = 0$ provides the final expression

$$y = \frac{lH_T}{w} \cosh\left(\frac{w}{lH_T}x\right). \qquad (3.54)$$

This is the answer we're after—a formula for the arc traced out by hanging chains (and power lines, ropes, etc.). By the symbols in (3.54), we have disproved the impression that hanging chains succumb to parabolas under gravity. In fact, their true shape is that of a hyperbolic trigonometric function, and solely depends upon the weight and length of the chain, as well as the horizontal tension exhibited by its end points. These arcs are referred to as *catenaries*, after the Latin word *catenarius* for anything that relates to a chain. Hopefully this will make for some fun and intelligent car-talk the next time you drive by some power lines.

§3.9 Buffon's Needle Problem

Buffon's Needle problem concerns, yet again, probability theory and π. Unlike our last encounter in this realm (§3.6), however, this section neither involves the normal distribution nor variants of Gaussian integrals. Instead we are concerned with the experiment posed in 1777 by the French nobleman Comte de Buffon who asked:

> *Suppose you drop a short needle on ruled paper. What is the probability that the needle settles into a position such that it crosses*

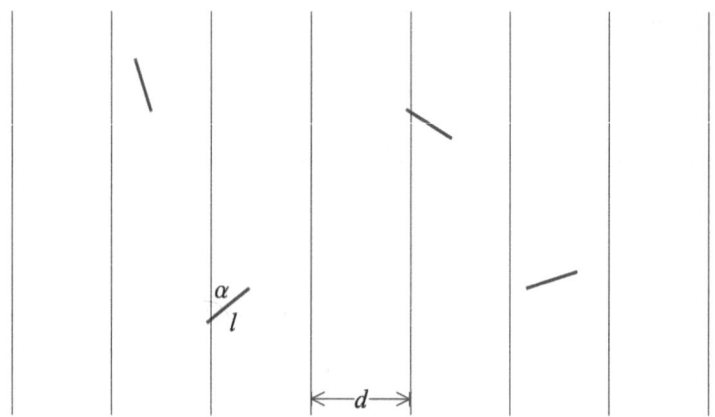

Figure 3.14: Buffon's Needles with short needles ($l \leq d$).

a line?

The probability depends on the relationship between the length of the needle l and the line separation distance d. This is visually obvious when comparing Figures 3.14 and 3.15 (page 112). Moreover, the probability depends on the angle α that the needle assumes relative to the vertical ruled lines. For $\alpha = \frac{\pi}{2}$, the needle has the greatest chance of crossing a line, while for $\alpha = 0$ the needle cannot cross a line (the probability that a needle lands exactly on and parallel to a line is zero, similar to how the dart thrower in §3.6 has a zero probability of striking any particular value on the dartboard). Due to angular symmetry, we are only concerned for $0 \leq \alpha \leq \frac{\pi}{2}$.

Let's now compute the probability for the short needle case ($l \leq d$), as in Figure 3.14. We can simplify the problem by first assuming the thrown needle always lands perpendicular to the ruled lines (this is $\alpha = \frac{\pi}{2}$). For $l \leq d$, the probability a horizontal needle crosses a line is $\frac{l}{d}$. Obviously, needles will not always land in this manner. This, however, is no problem since the horizontal distance $l\sin(\alpha)$ is completely analogous to a shorter, horizontal needle landing on

§3.9. Buffon's Needle Problem

the board. Therefore, for any thrown needle that lands at an angle α, we can simply imagine it as being a shorter needle that lands horizontally (perpendicular to the ruled lines) with length $l\sin(\alpha)$. By the same reasoning above, the probability of this shorter needle (and thus any needle) crossing a line is $\frac{l\sin(\alpha)}{d}$. Of course, we know nothing about the angle α, other than its range $0 \leq \alpha \leq \frac{\pi}{2}$ and the hypothesis that α is not favored towards any value within this spectrum. It is fair to assume, then, that all choices of α are random and independent between needles. It follows that the total probability p a needle crosses is the average of the probabilities over all possible angles α. That is,

$$p = \frac{1}{\frac{\pi}{2} - 0} \int_0^{\frac{\pi}{2}} \frac{l\sin(\alpha)}{d} d\alpha,$$

which evaluates to

$$p = \frac{2l}{\pi d}. \tag{3.55}$$

The interest in this probability is illuminated when we imagine actually performing Buffon's needle experiment. Suppose we set up the board as in Figure 3.14 with paper whose ruling distance $d \geq l$ (i.e. short needles). Onto this board we then toss many needles, say N needles, in a random fashion. The probability p that a needle crosses is clearly

$$p = \left(\frac{1}{N}\right) Np \tag{3.56}$$

Here the product Np is the expected number of needles that will cross. Hence, if

$$N_{\text{cross}} = Np$$

is the subset of the N needles that cross a line on the paper, it follows from (3.56) that the probability a needle crosses is

$$p = \frac{N_{\text{cross}}}{N}.$$

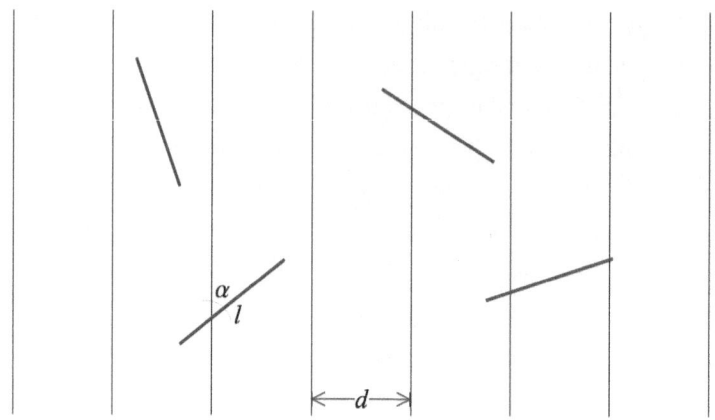

Figure 3.15: Buffon's Needles with long needles ($l > d$).

Equivalently, we derived in (3.55) that $p = \frac{2l}{\pi d}$. Thus, it must be that

$$\frac{N_{\text{cross}}}{N} = \frac{2l}{\pi d}.^\dagger$$

A simple rearrangement tells us that

$$\pi = \frac{2lN}{dN_{\text{cross}}}.$$

What we have just shown is that one can determine the value of π by tossing needles onto ruled paper. Indeed, you are encouraged to try this yourself.[‡]

Let's now consider the case of longer needles ($l > d$), as in Figure 3.15. We can intuit that the probability a long needle crosses is surly greater than for a short needle ($l \leq d$). To substantiate this intuition, we begin where we did with short needles—considering the horizontal distance $l \sin(\alpha)$ of a thrown needle.

[†]This is certainly not going to be exact for small N. But as $N \to \infty$ (as the number of needles increases) this becomes a genuine equality.

[‡]We obtained a fair result ($\pi \approx 3.1458$) with $d = 2l$ (so that $\pi = \frac{N}{N_{\text{cross}}}$) and $N = 151$.

§3.9. Buffon's Needle Problem

Notice that so long the horizontal distance is less than (or equal to) the ruling distance d—that is, so long $l\sin(\alpha) \leq d$—the problem is identical to the case for $l \leq d$ and the probability of crossing is simply $\frac{l\sin(\alpha)}{d}$, where $0 \leq \alpha \leq \arcsin\left(\frac{d}{l}\right)$. For $\alpha > \arcsin\left(\frac{d}{l}\right)$ the needle is guaranteed to cross because its length exceeds the ruling distance d. Ergo, for $\arcsin\left(\frac{d}{l}\right) < \alpha \leq \frac{\pi}{2}$ the probability the needle crosses is exactly one. Integrating as before, the total probability p is thus

$$p = \frac{1}{\frac{\pi}{2} - 0}\left(\int_0^{\arcsin(d/l)} \frac{l\sin(\alpha)}{d}\,d\alpha + \int_{\arcsin(d/l)}^{\frac{\pi}{2}} 1\,d\alpha\right)$$

$$= \frac{2}{\pi}\left[-\frac{l\cos(\alpha)}{d}\bigg|_0^{\arcsin(d/l)} + \frac{\pi}{2} - \arcsin\left(\frac{d}{l}\right)\right].$$

Evaluating, we obtain

$$p = 1 + \frac{2}{\pi}\left[\frac{l}{d} - \frac{l}{d}\cos\left(\arcsin\left(\frac{d}{l}\right)\right) - \arcsin\left(\frac{d}{l}\right)\right].$$

To compute $\cos\left(\arcsin\left(\frac{d}{l}\right)\right)$, take note that $\sin(\alpha) = \frac{d}{l}$. It follows from the trigonometric identity $\sin^2(\alpha) + \cos^2(\alpha) = 1$ that

$$\cos(\alpha) = \sqrt{1 - \frac{d^2}{l^2}}.$$

But $\alpha = \arcsin\left(\frac{d}{l}\right)$, hence $\cos\left(\arcsin\left(\frac{d}{l}\right)\right) = \sqrt{1 - \frac{d^2}{l^2}}$. Substituting this into the integral evaluation above, we conclude that

$$p = 1 + \frac{2}{\pi}\left[\frac{l}{d}\left(1 - \sqrt{1 - \frac{d^2}{l^2}}\right) - \arcsin\left(\frac{d}{l}\right)\right]. \qquad (3.57)$$

Observe in the limit $l \to d$ (long needles becoming short),

$$1 + \frac{2}{\pi}\left[\frac{l}{d}\left(1 - \sqrt{1 - \frac{d^2}{l^2}}\right) - \arcsin\left(\frac{d}{l}\right)\right]$$

$$\to 1 + \frac{2}{\pi}\left(\frac{l}{d} - \frac{\pi}{2}\right) = \frac{2l}{\pi d},$$

which is the same answer in (3.55). Though this answer is more complex than before, it does indeed simplify to the limit we expect with shorter needles. Moreover, observe that as $l \to \infty$, (3.57) tends toward one. We leave it to the reader to conclude why this makes sense.

§3.10　On the Irrationality of π

When posed with the task of providing an example of an irrational number, one's immediate instinct is to choose either $\sqrt{2}$ or π—these are the quintessential irrational numbers. We show in §A.1 that indeed $\sqrt{2} \notin \mathbb{Q}$. Yet in §1.1, we asserted $\pi \notin \mathbb{Q}$ without any justification for doing so. In this section we prove this to be true. Though before we begin, we acknowledge that this is the most sophisticated proof in the entire book. While it uses ideas well within our grasp, the conglomeration of these ideas is not immediately straightforward. There are steps in this proof whose motivation cannot be given due to their advanced nature—such as the rationale behind a few functions we seem to pull out of thin air. Nonetheless, the proof is well within our abilities provided we take it slow and step-by-step.

The proof begins with a word on notation and the function $f_n(x)$ defined by

$$f_n(x) := \frac{x^n}{n!}(1-x)^n.$$

With regard to notation, we will shortly be taking kth order derivatives of this function. Rather than using the traditional derivative notation $\frac{d^k f_n}{dx^k}$, we use a variant on prime notation:

$$f_n^{(k)} = \frac{d^k f_n}{dx^k}.$$

Make sure to not confuse the superscript with a power. (The rule is simple: If in parentheses, it is a derivative; otherwise, it is an exponent.)

§3.10. On the Irrationality of π

Our goal with $f_n(x)$ is to show $f_n^{(k)}(0)$ (the k derivative of f_n evaluated at $x = 0$) is an integer for all k, and subsequently that, assuming π is rational, establish an equality similar to $0 < f_n^{(k)}(0) < 1$. This inequality is surly absurd because there are no integers between zero and one. This is the contradiction we seek.

Looking at the definition of $f_n(x)$ it is clear that for $0 < x < 1$,

$$0 < f_n(x) < \frac{1}{n!}.$$

We now want to expand the product $x^n(1-x)^n$ in the definition of $f_n(x)$. We have

$$
\begin{aligned}
f_n(x) &= \frac{x^n}{n!}(1-x)^n \\
&= \frac{x^n}{n!}\underbrace{(1-x)(1-x)\cdots(1-x)}_{n \text{ factors}} \\
&= \frac{x^n}{n!}(1-2x+x^2)\underbrace{(1-x)(1-x)\cdots(1-x)}_{n-2 \text{ factors}} \\
&= \frac{x^n}{n!}(1-3x+3x^2-x^3)\underbrace{(1-x)(1-x)\cdots(1-x)}_{n-3 \text{ factors}}.
\end{aligned}
$$

Though we neglect the exact pattern, we are confident f_n assumes the form

$$f_n(x) = \frac{x^n}{n!}\left(1 + c_1 x + c_2 x^2 + \cdots + c_n x^n\right),$$

where all $c_1, c_2, \cdots, c_n$ are integers. Bringing in the factor of x^n, we obtain

$$f_n(x) = \frac{1}{n!}\left(x^n + c_1 x^{n+1} + c_2 x^{n+2} + \cdots + c_n x^{2n}\right). \quad (3.58)$$

We now compute $f_n^{(k)}(x)$ for $k > 2n$. Recognize that (3.58) is just a polynomial of degree $2n$. By the derivative rule

$$\frac{d}{dx} x^{2n} = 2n x^{2n-1}$$

any kth order derivative such that $k > 2n$ will be zero. Hence for $k > 2n$,
$$f_n^{(k)}(x) = 0 \implies f_n^{(k)}(0) = 0.$$

For $k < n$, notice that by repeated iterations of the derivative rule $\frac{d}{dx}x^n = nx^{n-1}$,
$$\frac{d^k}{dx^k}x^n = n \times (n-1) \times \cdots \times (n-(k+1))x^{n-k}$$
$$= \frac{n!}{(n-k)!}x^{n-k}$$

and so
$$f_n^{(k)}(x) = \frac{1}{n!}\left(\frac{n!}{(n-k)!}x^{n-k} + \frac{c_1(n+1)!}{(n+1-k)!}x^{n+1-k} + \cdots \right.$$
$$\left. + \frac{c_n(2n)!}{(2n-k)!}x^{2n-k}\right).$$

Therefore with $x = 0$ and $k < n$
$$f_n^{(k)}(0) = 0.$$

So far we have proven $f_n^{(k)}(0) = 0$ whenever $k < n$ and $k > 2n$. For the values in the middle, k is such that $n < k < 2n$. Using the fact that the nth derivative of an $(n-1)$th order polynomial is zero, the first $k-1$ terms in the sum
$$f_n^{(k)}(x) = \frac{d^k}{dx^k}\frac{1}{n!}\left(x^n + c_1 x^{n+1} + c_2 x^{n+2} + \cdots + c_n x^{2n}\right).$$

are zero. In other words, for $n < k < 2n$,
$$f_n^{(k)}(x) = \frac{1}{n!}\left(c_{k-n}k! + c_{k+1-n}\frac{(k+1)!}{1!}x + \cdots \right.$$
$$\left. + c_{2n}\frac{(2n)!}{(2n-k)!}x^{2n-k}\right).$$

Hence with $x = 0$ and $n < k < 2n$,
$$f_n^{(k)}(0) = c_{k-n}k!,$$

§3.10. On the Irrationality of π

which is clearly an integer. From a similar argument, we can rule out the particular cases where $k = n$ and $k = 2n$ in which we find $f_n^{(k)}(0) \in \mathbb{Z}$. Consequently,

> $f_n^{(k)}(0)$ is an integer for all k.

Notice that for the argument $1 - x$,

$$f_n(1-x) = \frac{(1-x)^n}{n!}(1-(1-x))^n = \frac{(1-x)^n}{n!}x^n = f_n(x).$$

Hence,

$$f_n(x) = f_n(1-x).$$

Taking the kth derivative of both sides, we have

$$f_n^{(k)}(x) = (-1)^k f_n^{(k)}(1-x).$$

With $x = 0$, this equation becomes

$$f_n^{(k)}(0) = (-1)^k f_n^{(k)}(1).$$

And based on our conclusion that $f_n^{(k)}(0) \in \mathbb{Z}$ for all k, it follows that $f_n^{(k)}(1)$ is also an integer for all k.

This concludes our preliminary analysis of $f_n(x)$. We now transition to prove the less-involved lemma that for large enough n, the ratio

$$\frac{a^n}{n!} < 1, \tag{3.59}$$

where the real number $a > 1$. We will see that this result establishes the contradiction. To prove (3.59) for sufficiently large n, we define the function

$$\chi(n) := \frac{a^n}{n!}.$$

Observe

$$\frac{\chi(n+1)}{\chi(n)} = \frac{a}{n+1}.$$

Hence, for sufficiently large n, the ratio

$$\frac{\chi(n+1)}{\chi(n)} < 1 \implies \chi(n+1) < \chi(n).$$

This means that after the threshold where n becomes sufficiently large, $\chi(n)$ is an always decreasing function. This implies the ratio $\frac{a^n}{n!}$ must be tending towards zero as $n \to \infty$. But to reach zero, it must first reach one. Therefore, for large enough n, we obtain the inequality in (3.59).

Okay, we are now set for the main proof. For sake of contradiction, we assume π^2 is rational. If this is so, then π is not necessarily rational.[†] However, if we manage to prove π^2 is irrational, then, necessarily, π is also irrational (why?). We will thus assume π^2 is rational and later unearth a contradiction, thereby proving the irrationality of π^2 and hence π.

Assuming π^2 is rational, there exist positive integers a and b such that

$$\pi^2 = \frac{a}{b}.$$

Define

$$G(x) := b^n \left[\pi^{2n} f_n(x) - \pi^{2n-2} f_n^{(2)}(x) + \pi^{2n-4} f_n^{(4)}(x) - \cdots \right. \\ \left. + (-1)^n f_n^{(2n)}(x) \right].$$

Here the magnitude of each term can be expressed generally as $b^n \pi^{2n-2k} f_n^{(k)}(x)$. And by our assumption $\pi^2 = \frac{a}{b}$, we have

$$\begin{aligned} b^n \pi^{2n-2k} f_n^{(k)}(x) &= b^n \pi^{2(n-k)} f_n^{(k)}(x) \\ &= b^n \left(\pi^2\right)^{n-k} f_n^{(k)}(x) \\ &= b^n \left(\frac{a}{b}\right)^{n-k} f_n^{(k)}(x) \\ &= b^k a^{n-k} f_n^{(k)}(x). \end{aligned}$$

[†]For example, $(\sqrt{2})^2$ is rational but $\sqrt{2}$ is irrational. Then again, 2^2 is rational and so is 2.

§3.10. On the Irrationality of π

Evidently, the factor $b^k a^{n-k} \in \mathbb{Z}$ for all k. And because $f_n^{(k)}(0)$ and $f_n^{(k)}(1)$ are integers for all k, we have shown yet another critical result:

> $G(0)$ and $G(1)$ are both integers.

The next step in this proof is to differentiate $G(x)$ twice, yielding

$$G''(x) = b^n \left[\pi^{2n} f_n^{(2)}(x) - \pi^{2n-2} f_n^{(4)}(x) + \cdots \right. \\ \left. + (-1)^n f_n^{(2n+2)}(x) \right].$$

However, by its definition $f_n(x)$ is only a $(2n)$th order polynomial. Hence, the $(2n+2)$th derivative of $f_n(x)$ is zero—that is,

$$(-1)^n f_n^{(2n+2)}(x) = 0.$$

It follows that

$$G''(x) = b^n \left[\pi^{2n} f_n^{(2)}(x) - \pi^{2n-2} f_n^{(4)}(x) + \cdots \right. \\ \left. + (-1)^{n-1} \pi^2 f_n^{(2n)}(x) \right].$$

If we now add $G''(x)$ and $\pi^2 G(x)$, we obtain the concise result

$$\begin{aligned} G''(x) + \pi^2 G(x) &= \pi^{2n+2} f_n(x) b^n \\ &= \pi^2 \left(\frac{a}{b}\right)^n f_n(x) b^n \\ &= \pi^2 f_n(x) a^n. \end{aligned} \qquad (3.60)$$

Now let

$$H(x) := G'(x) \sin(\pi x) - \pi G(x) \cos(\pi x),$$

whose first derivative follows from the product rule:

$$H'(x) = \pi G'(x)\cos(\pi x) + G''(x)\sin(\pi x) + \pi^2 G(x)\sin(\pi x)$$
$$- \pi G'(x)\cos(\pi x)$$
$$= \underbrace{\pi G'(x)\cos(\pi x) - \pi G'(x)\cos(\pi x)}_{\text{these terms vanish}}$$
$$+ G''(x)\sin(\pi x) + \pi^2 G(x)\sin(\pi x)$$
$$= \underbrace{[G''(x) + \pi^2 G(x)]}_{\text{this is (3.60)}}\sin(\pi x)$$
$$= \pi^2 f_n(x) a^n \sin(\pi x).$$

We now integrate as follows:

$$\int_0^1 \pi^2 f_n(x) a^n \sin(\pi x) dx = \int_0^1 H'(x) dx$$
$$= H(1) - H(0)$$
$$= -\pi G(1)\cos(\pi) + \pi G(0)\cos(0)$$
$$= \pi [G(1) + G(0)].$$

That is,

$$\int_0^1 \pi^2 f_n(x) a^n \sin(\pi x) dx = \pi [G(1) + G(0)]$$
$$\implies \int_0^1 \pi f_n(x) a^n \sin(\pi x) dx = G(1) + G(0).$$

And because $G(1)$ and $G(0)$ were previously shown to be integers, their sum is also an integer. Hence the integral

$$\int_0^1 \pi f_n(x) a^n \sin(\pi x) dx$$

must be an integer as well.

We are now almost there. Earlier we mentioned that for $0 < x < 1$,

$$0 < f_n(x) < \frac{1}{n!} \implies 0 < \pi f_n(x) a^n \sin(\pi x) < \frac{\pi a^n \sin(\pi x)}{n!}.$$

§3.10. On the Irrationality of π

Because $\sin(\pi x) \leq 1$ for all x, we have that

$$\frac{\pi a^n \sin(\pi x)}{n!} \leq \frac{\pi a^n}{n!}$$

and so

$$0 < \pi f_n(x) a^n \sin(\pi x) < \frac{\pi a^n}{n!}.$$

Integrating through the inequality, we obtain

$$\int_0^1 0 \, dx < \int_0^1 \pi f_n(x) a^n \sin(\pi x) \, dx < \int_0^1 \frac{\pi a^n}{n!} \, dx,$$

which evaluates to the following:

$$0 < \int_0^1 \pi f_n(x) a^n \sin(\pi x) \, dx < \frac{\pi a^n}{n!}.$$

Up to now, we have not provided any conditions on n. Hence, this inequality must be true for all n.

Now in (3.59) we proved the ratio $\frac{a^n}{n!}$ can be made less than one for sufficiently large n. This fact is true regardless of any scalar multiple of $\frac{a^n}{n!}$. Thus, for large enough n we will eventually have

$$0 < \int_0^1 \pi f_n(x) a^n \sin(\pi x) \, dx < 1.$$

Such is an impossible expression, for we recently deduced that the integral

$$\int_0^1 \pi f_n(x) a^n \sin(\pi x) \, dx \in \mathbb{Z}.$$

Yet there are no integers between zero and one, so we've reached a contradiction. This means our assumption that $\pi^2 \in \mathbb{Q}$ is incorrect, proving that π^2 is irrational. And from the discussion on page 118, this necessarily implies $\pi \notin \mathbb{Q}$—the result to be shown.

CHAPTER 4

SERIES AND PRODUCTS

"There is no division nor subtraction in the heart-arithmetic of a good mother. There are only addition and multiplication."
~ Bess Streeter Aldrich

EVERYTHING up to now has involved the foundational idea of a limit. In differential calculus we use limits to deduce the instantaneous slope of a line and in integral calculus they provide a means to compute the exact area under nonlinear curves. Our use of limits in this chapter is a new beast altogether. Here we traverse calculus over to its companion field *mathematical analysis* wherein we explore arithmetical operations, like addition and multiplication, to an infinite extent. Here limits help us discern to what value an infinite sum or product appears to be approaching as more terms are tacked on. Infinite operations like these are called *infinite series* for addition and *infinite products* for multiplication. As we will see, it is this perplexing notion of executing an operation an infinite number of times that leads to many of the unforeseen equalities in this chapter.

To begin, it is necessary to get a handle on series and product notation. We begin with a finite series that we call S and defined by the Greek conglomeration

$$S := \sum_{n=1}^{r} a_n. \tag{4.1}$$

Here the large Σ is the uppercase, Greek letter sigma and means "sum." (4.1) is equivalent to saying

$$S = a_1 + a_2 + a_3 \cdots + a_{r-1} + a_r,$$

where each a_n is a number defined by some expression. For example, if we state

$$T := \sum_{n=1}^{7} \frac{n^2}{(n+1)},$$

this is equivalent to the sum

$$T = \frac{1}{2} + \frac{4}{3} + \frac{9}{4} + \frac{16}{5} + \frac{25}{6} + \frac{36}{7} + \frac{49}{8}$$

because $a_n = \frac{n^2}{(n+1)}$, with n progressing from $n=1$ to $n=7$ in integer steps.

What we have just described is a *finite series* since the number of terms in the sum is finite. We call a series an *infinite series* in the limit

$$\lim_{r \to \infty} \sum_{n=1}^{r} a_n$$

where the number of terms increase without bound. Similar to improper integrals, to simplify notation we often omit the limit sign and just say

$$\lim_{r \to \infty} \sum_{n=1}^{r} a_n = \sum_{n=1}^{\infty} a_n.$$

At first, the idea of a sum with an infinite number of terms sounds preposterous. For starters, how is it possible to add an infinite amount of numbers without ever stopping and without the sum just becoming arbitrarily large? It turns out we can address both of these questions with a somewhat contrived example, the series

$$Z := \sum_{n=0}^{\infty} \frac{1}{2^n} = \frac{1}{2^0} + \frac{1}{2^1} + \frac{1}{2^2} + \frac{1}{2^3} + \cdots$$
$$= 1 + \frac{1}{2} + \frac{1}{2^2} + \frac{1}{2^3} + \cdots.$$

Taking $\frac{1}{2}Z$ we obtain

$$\frac{1}{2}Z = \frac{1}{2} + \frac{1}{2^2} + \frac{1}{2^3} + \frac{1}{2^4} + \cdots.$$

Hence,

$$Z - \frac{1}{2}Z = \left(1 + \frac{1}{2} + \frac{1}{2^2} + \cdots\right) - \left(\frac{1}{2} + \frac{1}{2^2} + \frac{1}{2^3} + \cdots\right).$$

All the terms cancel (with the exception of the lone one) leaving

$$\frac{1}{2}Z = 1 \implies Z = 2.$$

So even though we summed an infinite number of positive numbers, the value is finite and, even better, was computed in a finite amount of time.

The series Z above is part of a larger class of series called *infinite geometric series*. These are any series of the form

$$\sum_{n=0}^{\infty} C\omega^n = C\omega^0 + C\omega^1 + C\omega^2 + C\omega^3 + \cdots, [†] \quad (4.2)$$

where C and $\omega \in \mathbb{R}$. We show in §A.2 that if $|\omega| < 1$, then (4.2) *converges* to a definite, finite value defined by the simple fraction $\frac{C}{1-\omega}$. If instead $|\omega| \geq 1$, (4.2) does not approach any finite value and is therefore said to *diverge*.

We now transition our focus from addition (series) to multiplication (products). Similar to series, products are expressed succinctly

---

[†]You may recall this series from our discussion on the Koch snowflake in §2.5. Indeed, geometric series appear all over the place.

by the symbol Π, which is the capital Greek letter pi. A general product P is expressed in the following manner:

$$P := \prod_{n=1}^{r} a_n,$$

which is to be interpreted as the product

$$P = a_1 \times a_2 \times a_3 \times \cdots \times a_{r-1} \times a_r.$$

As with infinite series, an *infinite product* has an infinite number of factors:

$$\lim_{r \to \infty} \prod_{n=1}^{r} a_n = \prod_{n=1}^{\infty} a_n.$$

For instance, in defining

$$Q := \prod_{n=1}^{\infty} \frac{1}{n},$$

this is to be interpreted as the boundless product

$$Q = 1 \times \frac{1}{2} \times \frac{1}{3} \times \frac{1}{4} \times \cdots.$$

Like infinite series, we say Q *converges* if the product approaches a finite value and say that Q *diverges* if it does not.

Before diving into the meat of this chapter, we'll make three quick points about infinite products. First, recall the multiplication property of zero, which states that any number times zero is zero. Thus, if any factor in an infinite product is zero, the whole product equals zero, regardless of the other factors. Second, infinite products are really just infinite series in disguise. To see how, define

$$R := \prod_{n=1}^{\infty} a_n = a_1 \times a_2 \times a_3 \times \cdots.$$

Recalling the logarithm property $\log(mn) = \log(m) + \log(n)$, it follows that

$$\log(R) = \log(a_1 \times a_2 \times a_3 \times \cdots) = \log(a_1) + \log(a_2) + \log(a_3) + \cdots.$$

That is,
$$\log(R) = \log\left(\prod_{n=1}^{\infty} a_n\right) = \sum_{n=1}^{\infty} \log(a_n).$$
To put it another way,
$$\prod_{n=1}^{\infty} a_n = \exp\left(\sum_{n=1}^{\infty} \log(a_n)\right). \tag{4.3}$$

Though we will not manipulate infinite products in this way, this series-product relationship is one that should not be overlooked. (4.3) is often a valuable asset when proving the convergence or divergence of infinite products. Finally, a property of (some) products that we will use later in this chapter is their relationship to the factorial function. Recall that for an integer n, n-factorial is defined by the product

$$n! := \prod_{k=1}^{n} k = 1 \times 2 \times 3 \times \cdots \times (n-1) \times n.$$

Besides providing a short-hand for product notation, the product-factorial relationship is often exploited to prove the convergence or divergence of infinite products. For example, by the product above

$$Q := \prod_{n=1}^{\infty} \frac{1}{n} = 1 \times \frac{1}{2} \times \frac{1}{3} \times \frac{1}{4} \times \cdots,$$

it should be obvious that

$$Q = \lim_{n \to \infty} \frac{1}{n!},$$

which clearly converges to zero. Of course, not all products can be expressed using the factorial function. For instance, the finite product

$$\prod_{k=1°}^{89°} \tan(k) = \tan(1°) \times \tan(2°) \times \cdots \times \tan(89°)^\dagger \tag{4.4}$$

[†] Note that we are using degrees not radians here.

has no obvious relationship with the factorial function. Accordingly, its evaluation requires a separate technique. We could use the product-sum relationship in (4.3), but there is a more clever way to go about it. Notice (4.4) can be reformulated so that

$$\tan(1°) \times \tan(2°) \times \cdots \times \tan(89°) = [\tan(1°) \times \tan(89°)]$$
$$\times [\tan(2°) \times \tan(88°)] \times \cdots \times \tan(45°). \quad (4.5)$$

Here, each pair (excluding the sole $\tan(45°)$ factor) is of the form

$$\tan(x) \times \tan(90° - x) = \frac{\sin(x)}{\cos(x)} \times \frac{\sin(90° - x)}{\cos(90° - x)}.$$

Utilizing the trigonometric identities $\sin(x) = \cos(90° - x)$ and $\cos(x) = \sin(90° - x)$, it follows that $\tan(x) \times \tan(90° - x) = 1$ for all x. And because $\tan(45°) = 1$, we have

$$\prod_{k=1°}^{89°} \tan(k) = \underbrace{1 \times 1 \times \cdots \times 1}_{45 \text{ factors}} = 1.$$

On that note, we now have all the tools required to appreciate the results in this chapter. We'll start with one of the most famous series representations for π and work our way towards one of the greatest unsolved problems in all of mathematics.

§4.1 TAYLOR SERIES AND π

Our first journey into the fascinating world of series and products will be in exploring the first of four amazing formulae for π that will be encountered in this chapter. Before diving in, we need to address two special types of series: *Taylor* and *Maclaurin series*, named after the seventeenth-eighteenth century mathematicians Brook Taylor and Colin Maclaurin.

Formally, a *Taylor series* is a series expansion (i.e. representation) for a function $f(x)$. Functions and their Taylor series are analogous to fractions and their decimal expansions: While the fraction

$\frac{6}{7}$ is more abridged in this ratio representation, it is identical to the verbose decimal expansion $0.\overline{857142}$. By definition, the Taylor expansion of the function $f(x)$ is the series

$$f(x) = \sum_{n=0}^{\infty} \frac{f^{(n)}(c)(x-c)^n}{n!} \qquad (4.6)$$

$$= f(c) + f'(c)(x-c) + \frac{f''(c)(x-c)^2}{2!} + \cdots,$$

where $c \in \mathbb{R}$ and the notation $f^{(n)}(c)$ denotes the nth derivative of $f(x)$ at $x = c$. The special case with $c = 0$, which alters (4.6) so that

$$f(x) = \sum_{n=0}^{\infty} \frac{f^{(n)}(0)x^n}{n!} = f(0) + f'(0)x + \frac{f''(0)x^2}{2!} + \cdots, \qquad (4.7)$$

is called a *Maclaurin series*. For the remainder of this text we employ only the Maclaurin series representation for the function $f(x)$. Though while it is named differently, be cognizant that (4.7) is just a particular form of (4.6).

To grasp an understanding of how these expansions work, let's compute the Maclaurin series representations for the exponential, sine, and cosine functions. For $\exp(x)$, we know that $\exp(0) = 1$ and any order derivative of $\exp(x)$ is just $\exp(x)$. Therefore, for any $n \geq 0$

$$\frac{d^n}{dx^n}\exp(x) = \exp(x) \implies \left.\frac{d^n}{dx^n}\exp(x)\right|_{x=0} = \exp(0) = 1.$$

By (4.7), we have

$$\exp(x) = \sum_{n=0}^{\infty} \frac{x^n}{n!} = 1 + x + \frac{x^2}{2!} + \frac{x^3}{3!} + \frac{x^4}{4!} + \cdots. \qquad (4.8)$$

You are encourage to check that (4.8) satisfies all possible proceedings that $\exp(x)$ would pass, such as $\int_0^1 \exp(x)dx = e - 1.$[†] Unbelievably, (4.8) is a genuine equality.

[†]Somewhat circularly, this assumes you are aware of the identity

$$e = 1 + 1 + \frac{1}{2!} + \frac{1}{3!} + \frac{1}{4!} + \cdots. \qquad (4.9)$$

§4.1. Taylor Series and π

To compute the Maclaurin series for $\sin(x)$, we must first compute different order derivatives of $\sin(x)$. We can simplify this computation by acknowledging the repetitive behavior of trigonometric derivatives:

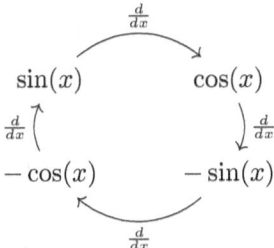

Here, each ⌢ denotes using the derivative operator $\frac{d}{dx}$ on the function at the node on the posterior end of each arrow. Reconciling this cyclic pattern with the derivatives evaluated at $x = 0$, the simple pattern emerges:

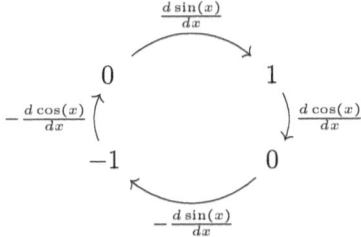

Assigning the proper value from this derivative orbit to that required by each term in (4.7), we find the Maclaurin series expansion of $\sin(x)$ to be

$$\sin(x) = \sum_{n=0}^{\infty} \frac{(-1)^n x^{2n+1}}{(2n+1)!} = x - \frac{x^3}{3!} + \frac{x^5}{5!} - \frac{x^7}{7!} + \frac{x^9}{9!} - \frac{x^{11}}{11!} + \cdots .$$

Though this follows from (4.8), (4.9) can be derived using the definition

$$e := \lim_{n \to \infty} \left(1 + \frac{1}{n}\right)^n.$$

See §A.5 for more details.

Similarly, for $\cos(x)$ we obtain

$$\cos(x) = \sum_{n=0}^{\infty} \frac{(-1)^n x^{2n}}{(2n)!} = 1 - \frac{x^2}{2!} + \frac{x^4}{4!} - \frac{x^6}{6!} + \frac{x^8}{8!} - \frac{x^{10}}{10!} + \cdots.$$

As with the exponential function, you are encouraged to play around with these to verify their equality with the sine and cosine functions.

The above expansions were merely examples for how series expansions are determined (that said, the above are integral to the next section). We now turn our attention to the somewhat arbitrary function $\arctan(x)$. As with the exponential, sine, and cosine functions, we wish to construct the Maclaurin expansion of $\arctan(x)$ and then, unlike the above functions, deduce a famous formulation for π.

We begin by computing a few derivatives of $\arctan(x)$, which are necessary to construct the Maclaurin expansion:

$$\frac{d}{dx} \arctan(x) = \frac{1}{1+x^2} \implies \frac{d}{dx} \arctan(x)\bigg|_{x=0} = 1.$$

The second derivative yields

$$\frac{d^2}{dx^2} \arctan(x) = -\frac{2x}{(1+x^2)^2} \implies \frac{d^2}{dx^2} \arctan(x)\bigg|_{x=0} = 0.$$

Taking higher order derivatives,[†] we find

$$\frac{d^3}{dx^3} \arctan(x)\bigg|_{x=0} = -2, \quad \frac{d^4}{dx^4} \arctan(x)\bigg|_{x=0} = 0,$$

$$\frac{d^5}{dx^5} \arctan(x)\bigg|_{x=0} = 24, \quad \cdots.$$

In general, if n is even then

$$\frac{d^n}{dx^n} \arctan(x)\bigg|_{x=0} = 0.$$

[†]These derivatives get quite messy so we've only incorporated them evaluated at $x = 0$. You are encouraged, of course, to justify our claims by evaluating these derivatives and deducing the general pattern.

If n is odd and is one *less* than a multiple of four, then

$$\frac{d^n}{dx^n} \arctan(x)\bigg|_{x=0} = -(n-1)!.$$

Otherwise, if n is odd and is one *greater* than a multiple of four, then

$$\frac{d^n}{dx^n} \arctan(x)\bigg|_{x=0} = (n-1)!.$$

Incorporating this trend into (4.7), we find

$$\arctan(x) = \sum_{n=0}^{\infty} \frac{(-1)^n x^{2n+1}}{2n+1} \tag{4.10}$$

$$= x - \frac{x^3}{3} + \frac{x^5}{5} - \frac{x^7}{7} + \frac{x^9}{9} - \frac{x^{11}}{11} + \cdots.$$

An important feature of Maclaurin series (and Taylor series in general) is that the trigonometric functions only satisfy them in radian units. This means that the fusion of $\arctan(1) = \frac{\pi}{4}$ and (4.10) corresponds to the endless sum

$$\sum_{n=0}^{\infty} \frac{(-1)^n}{2n+1} = 1 - \frac{1}{3} + \frac{1}{5} - \frac{1}{7} + \frac{1}{9} - \frac{1}{11} + \cdots = \frac{\pi}{4}. \tag{4.11}$$

This series, known as *Gregory's series* after the seventeenth century mathematician James Gregory, is perhaps the simplest infinite series for computing π. While profound, the rate at which (4.11) converges to $\frac{\pi}{4}$ is embarrassingly slow. Only after the first $500,000$ terms ($n = 500,000$) will (4.11) approximate π to five decimal places (that is, to the precision $\pi \approx 3.14159$). Despite our apparent impatience, it's impossible to not appreciate its beauty.

§4.2 Euler's Equation Revisited

In this section we again concern ourselves with Euler's prominent equation $e^{i\theta} = \cos(\theta) + i\sin(\theta)$ by going through a separate derivation using the Maclaurin polynomials derived in the previous section:

$$\exp(x) = e^x = \sum_{n=0}^{\infty} \frac{x^n}{n!} = 1 + x + \frac{x^2}{2!} + \frac{x^3}{3!} + \cdots \qquad (4.12)$$

$$\sin(x) = \sum_{n=0}^{\infty} \frac{(-1)^n x^{2n+1}}{(2n+1)!} = x - \frac{x^3}{3!} + \frac{x^5}{5!} - \frac{x^7}{7!} + \cdots \qquad (4.13)$$

$$\cos(x) = \sum_{n=0}^{\infty} \frac{(-1)^n x^{2n}}{(2n)!} = 1 - \frac{x^2}{2!} + \frac{x^4}{4!} - \frac{x^6}{6!} + \cdots . \qquad (4.14)$$

Unlike the derivation of Euler's equation in §3.2, our derivation here is quite straightforward and requires little insight.

We begin by considering (4.12) above. Suppose into $\exp(x)$ we substitute x such that $x = i\theta$ where $i := \sqrt{-1}$ is the imaginary unit. By (4.12), it follows that

$$e^{i\theta} = \sum_{n=0}^{\infty} \frac{(i\theta)^n}{n!} = 1 + i\theta + \frac{(i\theta)^2}{2!} + \frac{(i\theta)^3}{3!} + \frac{(i\theta)^4}{4!} + \cdots . \qquad (4.15)$$

Using the power properties of i—namely, $i^2 = -1$, $i^3 = -i$, $i^4 = 1$, and so forth—(4.15) becomes

$$e^{i\theta} = 1 + i\theta - \frac{\theta^2}{2!} - i\frac{\theta^3}{3!} + \frac{\theta^4}{4!} + i\frac{\theta^5}{5!} - \frac{\theta^6}{6!} - i\frac{\theta^7}{7!} + \cdots . \qquad (4.16)$$

Rearranging (4.16) into real and complex parts, we have

$$e^{i\theta} = \left(1 - \frac{\theta^2}{2!} + \frac{\theta^4}{4!} - \frac{\theta^6}{6!} + \cdots \right) + i\left(\theta - \frac{\theta^3}{3!} + \frac{\theta^5}{5!} - \frac{\theta^7}{7!} + \cdots \right).$$

That is,

$$e^{i\theta} = \left(\sum_{n=0}^{\infty} \frac{(-1)^n \theta^{2n}}{(2n)!}\right) + i\left(\sum_{n=0}^{\infty} \frac{(-1)^n \theta^{2n+1}}{(2n+1)!}\right). \qquad (4.17)$$

But the sum in the first term of (4.17) is identical to that in (4.14), which is $\cos(\theta)$. Likewise, the second term is identical to that in (4.13), which is $\sin(\theta)$. Hence, we've arrived at Euler's equation:

$$e^{i\theta} = \cos(\theta) + i\sin(\theta). \qquad (4.18)$$

§4.2. Euler's Equation Revisited

In §3.2, we pursued the interesting corollary of (4.18) that upon substituting $\theta = \pi$ we obtain the remarkable expression

$$e^{i\pi} = -1. \tag{4.19}$$

This, as you may recall, is Euler's identity. Here we would like to expand upon (4.19) to derive something new with logarithms. By applying the natural logarithm function on both sides of (4.19), we obtain the elegant expression

$$\log(-1) = i\pi.$$

By the logarithm property $\log(ab) = \log(a) + \log(b)$, it follows that the logarithm of any negative number $a \in \mathbb{R}$ is equivalent to

$$\begin{aligned}\log(-a) &= \log(-1 \times a) \\ &= \log(-1) + \log(a) \\ &= i\pi + \log(a).\end{aligned} \tag{4.20}$$

It seems that the pesky domain for $\log(x)$ being $x \in (0, \infty)$ can now be extended to $x \in (-\infty, 0) \cup (0, \infty)$[†]—the entire number line, with the exception of zero (there is no complex number z such that $e^z = 0$).

The astute reader will recall from §3.2 that, while $e^{i\pi} = -1$, so does $e^{i(3\pi)}$, $e^{i(5\pi)}$, $e^{i(7\pi)}$, and so forth. In fact, for any odd integer k

$$e^{i(k\pi)} = -1. \tag{4.21}$$

This, of course, is readily verified using (4.18). But this tells us something interesting about taking the logarithm of negative numbers—they too can take on many different values. To see this, note from (4.21) that

$$\log(-1) = ik\pi.$$

[†]The symbol $\cup$ is the set union symbol. This simply concatenates the two intervals.

The case $\log(-1) = i\pi$ is just the special case where $k = 1$. Together with (4.20), the general formula for a logarithm with argument $-a$ is thus

$$\log(-a) = ik\pi + \log(a).$$

In order to make results consistent, mathematicians specify the *principal branch* of the logarithm function. This is essentially a limit on the domain of the value k, restricting it so that the imaginary part of $\log(-1)$ is always within the range $(-\pi, \pi]$. The same is true of the remarkable

$$i^i = e^{-(4k+1)\pi/2}$$

encountered in §3.2. k can be restricted so that i^i doesn't take on infinitely many values, but just one value called its *principle value*—namely, $i^i \approx 0.20788$ ($k = 0$).

These are the kinds of analyses ubiquitous throughout the field of *complex analysis*. They are certainly unfamiliar, yet enticing nevertheless.

§4.3 The Harmonic Series

Among the infinite number of conceivable infinite series, the harmonic series is quite the celebrity. Its properties have many scratching their heads, especially those new to its peculiarities.

The infinite *harmonic series* is defined by the sum

$$H_\infty := \sum_{n=1}^{\infty} \frac{1}{n} = 1 + \frac{1}{2} + \frac{1}{3} + \frac{1}{4} + \cdots. \qquad (4.22)$$

Here the notation H_∞ is used so that a finitely-termed version of (4.22) is easily referenced:

$$H_m := \sum_{n=1}^{m} \frac{1}{n} = 1 + \frac{1}{2} + \frac{1}{3} + \frac{1}{4} + \cdots + \frac{1}{m}.$$

§4.3. THE HARMONIC SERIES

Mathematicians call H_m the mth *harmonic number*. Evidently, as $m \to \infty$, $H_m \to H_\infty$. This is (4.22), which is what we will focus on in the mean time. Towards the end of this section, however, we will transition our attention back to H_m and formulate one of the great unsolved problems in mathematics.

When encountering a new infinite series, it is only natural to wonder if the series converges. If so, it is then routine to deduce the value to which the series converges. For the harmonic series, we notice each successive term is smaller than the one preceding it, so it's feasible H_∞ converges to a finite value, which is what we would like to find. Such reasoning is superficially substantiated by noting the limit

$$\lim_{n \to \infty} \frac{1}{n} = 0,$$

implying the terms in the harmonic series approach zero, and so, at some point along in the summation, the series just gains negligible value in relation to the total value of the series at that point along in the adding. This, of course, is (near) the limit of the series. Associated with this reasoning, there is in fact a closely related theorem, which we prove below due to its serviceable simplicity:

Theorem 4.1. *If S is a convergent series defined by*

$$S := \sum_{n=1}^{\infty} a_n,$$

then as $n \to \infty$, $a_n \to 0$.

Proof. The proof is quite straightforward. Because S is convergent, it approaches some finite value L. After some arbitrary number of terms, say k terms, the series S will have some value V_k (this is like the mth harmonic number, except for the arbitrary series S). The value V_k is called the kth *partial sum* of S. Because S converges to L, so does the limit

$$\lim_{k \to \infty} V_k = L. \tag{4.23}$$

Of course, the same analysis applies to the $(k-1)$th partial sum:

$$\lim_{k\to\infty} V_{k-1} = L. \tag{4.24}$$

The partial sums V_k and V_{k-1} are related by the simple formula

$$V_k = a_k + V_{k-1},$$

which follows directly from the definition that V_k is the sum of all a_n in S up to the value a_k (where $n = k$). It follows from both (4.23) and (4.24) that

$$\lim_{k\to\infty} V_k = \lim_{k\to\infty} (a_k + V_{k-1})$$
$$L = \lim_{k\to\infty} (a_k) + L.$$

This is only possible if

$$\lim_{k\to\infty} a_k = 0.$$

Substituting n in for k completes the proof. □

We saw previously that the limit

$$\lim_{n\to\infty} a_n = 0 \tag{4.25}$$

applies to the harmonic series. Unfortunately, due to the conditionals in Theorem 4.1, such a limit does not tell us if H_∞ converges. While (4.25) is true of all convergent series, not all series for which (4.25) is true are convergent (read carefully into the conditionals of the theorem). Indeed, H_∞ is one such rebellious sum and actually diverges.

To prove this reasonably counterintuitive claim, we employ a separate method for determining whether a series converges or diverges: the *comparison test*. This evaluation considers two series at a time, one of which is the series in question (call this series A, for

§4.3. THE HARMONIC SERIES

us it's H_∞) and the other is a series familiar to us (that is, we know whether it converges or diverges—call this series B).

If testing for convergence, we require a convergent series B. If each term in B is greater than or equal to the corresponding term in A, it follows that A converges. This is valid because B essentially bounds the series A from above. If summing successive terms in A is thought of like the successive addition of water capable of being pumped into a water balloon, then B is like the sequential stretching of the balloon's elastic surface—it stretches for a while (increasing in value), but soon it reaches its limit and the balloon pops. This point is that at which the balloon can no longer be filled with water, and so A too has reached its limit.[†] So if indeed B converges (meaning the upper bound on A is finite), A must also converge.

Conversely, if testing for divergence you require a divergent series B. If each term in B is less than or equal to the corresponding term in A, it follows that A is also divergent. Here, rather than bounding from above, B acts to persuade A from below. This is analogous to hunger provoking one to buy more food than they need at the grocery store. If A is the number of items in your grocery basket and B is a measure of your insatiable hunger, B acts to increase A as time passes (as more terms in each series are added). Because your hunger B is unquenchable, you require an infinite amount of food to be satisfied. Hence, B forces A to be infinite as well, meaning A diverges in the same manner as B.

Using this idea, we can deduce the divergence of

$$H_\infty := 1 + \frac{1}{2} + \frac{1}{3} + \frac{1}{4} + \frac{1}{5} + \frac{1}{6} + \frac{1}{7} + \frac{1}{8} + \frac{1}{9} + \cdots.$$

[†]Note that the amount by which the surface of the balloon stretches and the amount of water added to the balloon before eruption will not necessarily be the same, similar to how the two limits of the series A and B need not be the same. The point of the analogy is to illustrate that they both have limits—that both A and B converge.

To do so, we introduce our series B:

$$B := 1 + \frac{1}{2} + \frac{1}{4} + \frac{1}{4} + \frac{1}{8} + \frac{1}{8} + \frac{1}{8} + \frac{1}{8} + \frac{1}{16} + \cdots.$$

We derived B by parsing through H_∞ and, for each term, computing the smallest power of a half less than or equal to it.[†] Observe that for the terms in B with power $\frac{1}{2^n}$, there exist 2^{n-1} terms with this value. For example, for the power $\frac{1}{2^2} = \frac{1}{4}$, there are two terms. This occurs because the difference between consecutive powers of two is itself a power of two. So between the last term in B with power $\frac{1}{2^{n-1}}$ and the last term in B with power $\frac{1}{2^n}$, there are

$$2^n - 2^{n-1} = 2^{n-1}(2-1) = 2^{n-1}$$

terms with power $\frac{1}{2^n}$. Thus, the amount these $\frac{1}{2^n}$ terms contribute to the sum is $2^{n-1}(\frac{1}{2^n}) = \frac{1}{2}$. And this is true for all n. Hence, B is equivalent to an infinite sum of halves:

$$B = 1 + \frac{1}{2} + \frac{1}{2} + \frac{1}{2} + \frac{1}{2} + \frac{1}{2} + \cdots.$$

In separate notation,

$$B = 1 + \lim_{p \to \infty} \frac{p}{2}.$$

This clearly diverges. And because every term in B is less than or equal to the corresponding term in the harmonic series, H_∞ must also diverge. Consequently, the harmonic series is a divergent sum. This should be surprising given Theorem 4.1 and just the simple structure of H_∞. Nonetheless, our justification here is well-founded.

We expressed above that a natural corollary to discovering a new convergent sum is in finding the value to which that series converges. Of course, this is inapplicable to divergent sums. What is

---

[†]For example, if the term in H_∞ is $\frac{1}{10}$, the smallest power of a half less than or equal to $\frac{1}{10}$ is $\frac{1}{16}$, so we add this in place of $\frac{1}{10}$ and continue for the remainder of the terms.

§4.3. The Harmonic Series

applicable, however, and often of interest is the rate of divergence—that is, how fast the series appears to diverge. Some are quite fast, such as the series

$$F := \sum_{n=1}^{\infty} (n^n)! = (1^1)! + (2^2)! + (3^3)! + (4^4)! + \cdots.$$

To get an idea, it takes F only three terms ($n = 3$) to reach the value

$$10,888,869,450,418,352,160,768,000,025,$$

roughly ten million million billion. By comparison, the harmonic series takes precisely $272,400,600$ terms ($n = 272,400,600$) to obtain a sum that exceeds 20. In other words, $H_{272,400,600} \approx 20$. Even more invigorating, for the harmonic series to surpass a value of 100 requires

$$15,092,688,622,113,788,323,$$
$$693,563,264,538,101,449,859,497$$

terms. That's approximately fifteen million million million million million terms! In relation to the age of the universe (roughly 13.7 billion years), the number above would have to be expressed in units of *yoctoseconds* (10^{-24}s—a septillionth of a second)[†] to have a value even remotely close to the value corresponding to the age of the universe in *seconds*.

There is little need for us to stress this implausibly slow pace any further. Yet there is another point worth noting about the growth of H_m. Though not necessary here, it can be shown that H_m grows alongside the natural logarithm function $\log(m)$, particularly as $m \to \infty$ (see Figure 4.1). In fact, as $m \to \infty$ the difference $H_m - \log(m)$ approaches a finite value γ called the *Euler-Mascheroni constant* after both Euler and the eighteenth century mathematician Lorenzo

---

[†]In one yoctosecond, light will travel approximately $(3 \times 10^8) \times 10^{-24} = 3 \times 10^{-16}$ meters—a distance less than the diameter of a proton.

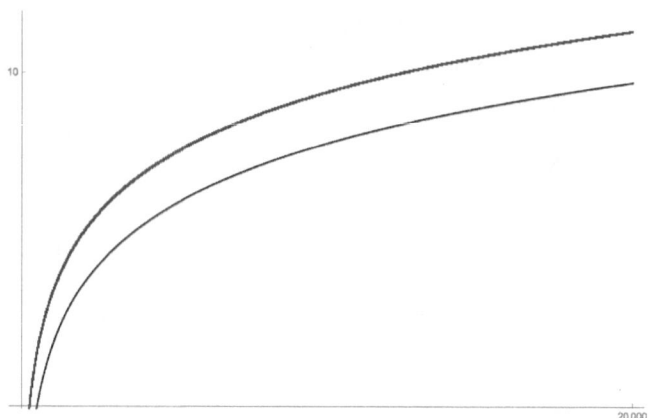

Figure 4.1: H_m (top curve) versus $\log(m)$ (lower curve) with $m \in [1, 20\,000]$.

Mascheroni. γ is defined by the limit

$$\gamma := \lim_{m \to \infty} H_m - \log(m) = \lim_{m \to \infty} \sum_{n=1}^{m} \frac{1}{n} - \log(m) = 0.577215\cdots.$$

As it turns out, very little is known about γ despite its omnipresence in both mathematics and physics. In fact, as we'll discuss in §5.7, mathematicians don't even know if γ is rational or not. While most suspect $\gamma \notin \mathbb{Q}$, this remains at the level of conjecture.

§4.4 THE BIRTHDAY PARADOX

This section represents the inaugural use of product notation in application to a surprising mathematics problem. We begin with the question we intend to answer:

> How many people must occupy a room so that the probability any two share a birthday is greater than 50%?

With such a short time to process the problem, your intuition may support the following argument. Try to spot the logical error:

§4.4. The Birthday Paradox

1. Let the number of people in the room be n, where $n \geq 2$.[†]

2. There are 365 days in a year. Suppose Josh, a person in the room, has a birthday on one of these days, say day X.

3. The probability another person in the room, Bujji, has a birthday on X is $P = \frac{1}{365}$.

4. Add another person to the room, Sharon, whose probability of sharing a birthday with Josh is also $\frac{1}{365}$. Adding an additional person, however, increases the total probability of *someone* (i.e. either Bujji or Sharon) sharing the birthday on X with Josh to $P = \frac{1}{365} + \frac{1}{365} = \frac{2}{365}$.

5. In general, for a room with n people, the probability of someone sharing a birthday with Josh is

$$P_{\text{Josh}} = \frac{n-1}{365}. \qquad (4.26)$$

6. We want to find n such that $P_{\text{Josh}} = 0.5$. It follows immediately from (4.26) that such a P_{Josh} is obtained when $n = 183.5$. Evidently, we cannot have half of a person, so we round up and conclude the room must house 184 people.

Did you weave out the error? If not, there is no need to worry. Perhaps the task "spot the logical error" purported the above enumeration would be an illogical argument. In fact, the argument above is perfectly logical, but is only useful in the context of the different problem:

> How many people must occupy a room so that the probability someone in the room shares a birthday with Josh (a person in

---

[†] We are interested in comparing pairs of people, so there are necessarily two or more people in the room.

the room) is greater than 50%?

Hence, it's not that the argument is illogical, in fact quite the contrary, we have simply failed at approaching the problem we are interested in. Our mistake lies in the fact that we only compared birthdays to a specific person in the room (namely Josh) and not birthdays *between* every person in the room. If we were to construct a solution around not just one person sharing a birthday with everyone else, but between all possible pairs of people in the room, it is reasonable to suspect the answer should be less (perchance much less) than 184 people.

To solve the problem correctly, it is of interest to first unearth an underlying principle in the problem—the *pigeonhole principle*. In its rudimentary form, the pigeonhole principle states that given n closed boxes and m rubber balls, if $m < n$ (more boxes than balls) it is possible to place each ball in its own box. However, if $m > n$ (more balls than boxes) then upon placing all the balls in the boxes, there must exist one box with at least two balls.[†]

Though this principle is intuitively obvious, it can lead to some startling conclusions. A favorite is that there exist at least two people in Boston, Massachusetts with the same number of hairs on their head. The average number of hairs on a human head is $150,000$ and the population of Boston is approximately $600,000$. Given the average number of hairs, it is safe to assume that no person in Boston has more than $500,000$ hairs on their head. We now assign a box (pigeonhole) to each number of hairs (to each value in the range 1 - $500,000$) and imagine the residents of Boston as rubber balls (pigeons). Since there are more residents than possible number of hairs (more balls than boxes; more pigeons than pigeonholes), it follows

---

[†]By now you are wondering why this is called the *pigeonhole* principle. To retrieve the original syntax of the idea, replace rubber balls by pigeons and boxes by pigeonholes. We amended our description because routine people (like the author) don't know what the heck a pigeonhole is.

§4.4. THE BIRTHDAY PARADOX

from the pigeonhole principle that at least two people have the same number of hairs on their head. Naturally, however, this principle is useful beyond counting hairs.

The pigeonhole principle allows us to organize the birthday problem in a way that is intuitively obvious and also rigorous: If the people in the room are the rubber balls (pigeons), each day of the year (ranging from 1 - 365) is a box (pigeonhole), and n is the number of people in the room, then for $n > 365$ the pigeonhole principle guarantees at least two people share a birthday.[†] It follows immediately that for $n > 365$ the probability at least two people share a birthday is 100%. In contrasting light, we can say that for $n > 365$ the probability no two people share a birthday is exactly zero.

Adjusting the number of people so that $n < 365$ implies that there are now more boxes than rubber balls, and so among all people in the room, two need not share a birthday, and so the probability no two people share a birthday is greater than zero. It turns out viewing the original birthday question from this negative lens (i.e. probabilities of things *not* occurring) is the best way to approach the problem. So to acquire the answer, let's proceed by first addressing the question:

> What is the probability that among n people ($n \geq 2$), Josh, a person in the room, does not share a birthday with anyone else?

On the previous page, we derived that among n people, the probability Josh shares a birthday with someone is $P_{\text{Josh}} = \frac{n-1}{365}$. Since probabilities must sum to one, the logical inverse of this statement (the question above) has a probability

$$Q_{\text{Josh}} = 1 - \frac{n-1}{365}. \qquad (4.27)$$

[†]This assumes a common 365 day year and that each of the n people have a birthday in the common year.

This is the probability of Josh not sharing a birthday with any of the n people in the room. We can perform a similar analysis for a different person, say Sharon. Assuming Josh does not share a birthday with anyone in the room, we automatically know Sharon does not share a birthday with Josh. Hence, we need not consider Josh in the calculation for Sharon since we are certain Josh and her birthday fail to coincide. Thus, we can treat the room as having $n-1$ people, making Sharon's probability of not sharing a birthday with anyone in the room

$$Q_{\text{Sharon}} = 1 - \frac{n-2}{365}. \qquad (4.28)$$

Put together, the probability both Josh and Sharon do not share a birthday with anyone in the room is the product of (4.27) and (4.28). And this is true in general: To compare between all possible pairs of people in the room, just multiply together the probabilities of people not sharing birthdays with one another, as in

$$Q(n) = \prod_{k=1}^{n-1} \left(1 - \frac{k}{365}\right). \qquad (4.29)$$

This is the probability that among n people, none of them share a birthday. This is the inverse of the probability we are interested in, so you may be inclined to take $1 - Q(n)$. Though because we want n such that $P(n) = 0.5$, by the equation $P(n) + Q(n) = 1$ it suffices to find the value of n such that $Q(n) = 0.5$. By (4.29), this value is obtained by multiplying

$$Q(n) = \left(1 - \frac{1}{365}\right) \times \left(1 - \frac{2}{365}\right) \times \cdots \times \left(1 - \frac{(n-1)}{365}\right)$$

until $Q(n) \geq 0.5$. To our astonishment, this is achieved when $n = 23$. Ergo, in a room of only 23 people, there is greater than a 50% chance that two share a birthday.

The shock in such a small number leaves some in complete disbelief. This is why the problem is called the *birthday paradox*—it

§4.5. ON THE INFINITUDE OF PRIMES

seems impossible for the probability to be so high with so few people. Yet the statistics speaks for itself. Of course, what is not paradoxical is the expectation that with more people in the room, the probability two share a birthday only increases—indeed, this is reflected by (4.29). Such is particularly true if twins are in the room—a situation whose analysis is left to the reader.

§4.5 ON THE INFINITUDE OF PRIMES

Likely the most pure form of inquiry in all mathematics is in the study of numbers, i.e. the field of *number theory*. And among all forms of numbers pursued in number theory, the most celebrated are the primes—the set $\mathbb{P}$.

As a reminder, a number $p \in \mathbb{N}$ is *prime* if, and only if, the divisors of p are 1 and p. Otherwise, p is a *composite number*. Number theorists will write $d \mid p$ to signify that d divides the number p. In other words, that there exists a natural number k such that $p = kd$. Likewise, the notation $d \nmid p$ articulates that there is no number k such that $p = kd$, meaning d does not divide p. Using this notation, if $d \mid p$, then p is prime only if $d = 1$ or $d = p$. If there exists a d for which this is not the case, then p is composite.

We prove in §A.6 that every composite number can be written as a unique product of primes. This, as you may know, is the *fundamental theorem of arithmetic*. Thus, due to the fact that there are infinitely many composite numbers (this is seen by multiplying 1×2, then $1 \times 2 \times 3$, then $1 \times 2 \times 3 \times 4$ and so forth), it's reasonable to suspect infinitely many primes since each composite number is a *unique* product of primes. It is the purpose of this section to prove the infinitude of primes using two separate formulations. While both methods are quite similar, the second is substantially shorter than the other.

In each, we utilize a method of proof called *reductio ad absurdum*

(Latin for "reduce to absurdity"). In more colloquial terms, this is a *proof by contradiction*. While the contents of these proofs will differ from case to case, the underlying structure is largely the same: Assume the opposite of the theorem, and show the assumption leads to a contradiction (such as $1 = 2$). In the following proofs, we assume the number of primes to be finite. We'll start with a proof by Euclid:

Theorem 4.2. *There are infinitely many prime numbers.*

Proof. Assume there are only finitely many prime numbers. If so, then the set of primes $\mathbb{P}$ has a finite number of elements. This means we can write out $\mathbb{P}$:

$$\mathbb{P} := \{p_1, p_2, p_3, \cdots, p_n\},$$

where $p_1, p_2, \cdots, p_n$ are all the distinct prime numbers, with p_n being the largest of such primes. Define

$$Q := 1 + \prod_{k=1}^{n} p_k = 1 + p_1 \times p_2 \times p_3 \times \cdots \times p_n$$

as one more than the product of all the prime numbers. Two properties are immediately obvious: $Q > p_n$ and Q is either prime or composite.

- If Q is prime, then we have found a prime not in the set $\mathbb{P}$. This means that $\mathbb{P}$ is incomplete. But this contradicts our assumption that $\mathbb{P}$ is composed of all the primes. Hence, there must be infinitely many primes.

- If Q is composite, then by the fundamental theorem of arithmetic there exists at least one prime $p_j \in \mathbb{P}$ such that $p_j \mid Q$. That is, there exists some $k \in \mathbb{N}$ such that

$$(1 + p_1 \times p_2 \times \cdots \times p_j \times \cdots \times p_n) = kp_j.$$

§4.5. On the Infinitude of Primes

Solving for k we obtain

$$k = \frac{1}{p_j} + p_1 \times p_2 \times \cdots \times p_{j-1} \times p_{j+1} \times \cdots \times p_n.$$

Because $p_j > 1$ (one is not prime, see §A.6), k is not an integer. Thus, $p_j \nmid Q$. It follows that $\mathbb{P}$ is incomplete because it does not contain the prime numbers that makeup Q. As before, this is a contradiction. Hence, there must be infinitely many primes.

In either case, the infinitude of primes is required. Thus, the theorem is proved. $\square$

For obvious reasons, this proof is credited as one of the more beautiful proofs in mathematics: It's relatively straightforward, not too complex, and quite profound. But a recent article published in the American Mathematical Monthly clearly outperforms Euclid's proof in both concision and grace.[†] This much shorter proof goes like this:

Theorem 4.3. *There are infinitely many prime numbers.*

Proof. Suppose there are only finitely many primes and let Q be their product; then

$$0 < \prod_{k=1}^{n} \sin\left(\frac{\pi}{p_k}\right) = \prod_{k=1}^{n} \sin\left(\frac{\pi(1+2Q)}{p_k}\right) = 0. \tag{4.30}$$

$\square$

Why this proves the theorem is our focus for the remainder of the section. A careful analysis reveals that there is indeed a contradiction in (4.30)—namely, the statement $0 < 0$. But how the proof arrives at this assertion is currently unknown. To understand why this esoteric formulation is correct, we will parse (4.30) from left to

---

[†] For the original article, see [16] in the bibliography.

right and justify why each relational operator (these are $<$, $=$, etc.) is right.

Consider first the assertion

$$0 < \prod_{k=1}^{n} \sin\left(\frac{\pi}{p_k}\right), \qquad (4.31)$$

where all p_k are distinct primes and p_n is the largest of such primes. As we know, the smallest prime number $p_1 = 2$, from which it follows that $\sin\left(\frac{\pi}{p_1}\right) = \sin\left(\frac{\pi}{2}\right) = 1$. All other primes are greater than two, hence

$$0 < \frac{\pi}{p_k} < \frac{\pi}{2} \implies 0 < \sin\left(\frac{\pi}{p_k}\right) < 1$$

for $k > 1$. Therefore, so long the product in (4.31) has only a finite number of terms (that is, so long the number of primes is finite), (4.31) will multiply together n positive factors producing both a nonzero and nonnegative product.

We now turn our attention to the statement

$$\prod_{k=1}^{n} \sin\left(\frac{\pi}{p_k}\right) = \prod_{k=1}^{n} \sin\left(\frac{\pi(1+2Q)}{p_k}\right), \qquad (4.32)$$

where $Q := p_1 \times p_2 \times \cdots \times p_n$ is the product of all the distinct prime numbers. Consider first the argument in the sine function on the right-hand side of (4.32). As we'll prove in §4.10, for real numbers A and B

$$\sin(A+B) = \sin(A)\cos(B) + \sin(B)\cos(A).$$

Employing this formula, (4.32) becomes

$$\sin\left(\frac{\pi(1+2Q)}{p_k}\right) = \sin\left(\frac{\pi}{p_k}\right)\cos\left(\frac{2\pi Q}{p_k}\right) \\ + \sin\left(\frac{2\pi Q}{p_k}\right)\cos\left(\frac{\pi}{p_k}\right). \qquad (4.33)$$

§4.5. On the Infinitude of Primes

Because p_k is a prime factor of Q, $p_k \mid Q$. Therefore, $\sin\left(\frac{2\pi Q}{p_k}\right) = 0$ and $\cos\left(\frac{2\pi Q}{p_k}\right) = 1$ as $\sin(2\pi\alpha) = 0$ and $\cos(2\pi\alpha) = 1$ for $\alpha = \frac{Q}{p_k} \in \mathbb{Z}$. This is true of all k because all p_k divide Q. In other words, we have from (4.33) that for all k

$$\sin\left(\frac{\pi(1+2Q)}{p_k}\right) = \sin\left(\frac{\pi}{p_k}\right).$$

Taking the product over all primes on each side of this expression constructs (4.32).

The final statement in (4.30) declares

$$\prod_{k=1}^{n} \sin\left(\frac{\pi(1+2Q)}{p_k}\right) = 0. \tag{4.34}$$

To substantiate this, we utilize our assumption on the finitude of primes. We acknowledge that the integer $1 + 2Q$ cannot itself be prime, else Q is not the product of all the primes (a contradiction). Therefore, $1 + 2Q$ must be composite and composed of some (if not all) of the primes that makeup Q. Though if this is true, then there exists at least one prime p_j that evenly divides $1 + 2Q$. Hence, when $k = j$ in (4.34),

$$\sin\left(\frac{\pi(1+2Q)}{p_j}\right) = 0, \tag{4.35}$$

because the quantity $\frac{1+2Q}{p_j} \in \mathbb{N}$. Now because (4.35) is zero and is one of n factors in (4.34), the whole product in (4.34) must equal zero.

Concatenating (4.31), (4.32), and (4.34) produces the net relation:

$$0 < \prod_{k=1}^{n} \sin\left(\frac{\pi}{p_k}\right) = \prod_{k=1}^{n} \sin\left(\frac{\pi(1+2Q)}{p_k}\right) = 0.$$

But this implies $0 < 0$—an obvious falsehood. The culprit inevitably lies in our assumption that there are only finitely many primes. We are thus forced to conclude with the logical inverse—that there are infinitely many primes.

While both of these proofs have their merits, it is impossible to not expatiate on one respectable conclusion: Even under the constant influence and persistent use of numbers that venture hardly beyond zero, mathematics allows us to grasp well-beyond our numerical comprehensions to prove that some entities are as vast as infinity—a notion no man has ever truly grasped. Quite remarkable, to say the least.

§4.6   On the Irrationality of e

Earlier in §1.2 we proved that the nth root of two is irrational for all $n > 2$. In §A.1 we extend this result to $n \geq 2$, and in §3.10 we proved that π is also irrational. In this section, we prove Euler's number
$$e := \lim_{n \to \infty} \left(1 + \frac{1}{n}\right)^n = 2.7182818\cdots$$
also belongs to the list of irrationals. The difficulty of this proof is in between that for $\sqrt[n]{2}$ and π. Nevertheless, it is well within our grasp.

The proof begins by recalling the series expansion of e^x:
$$e^x = \sum_{n=0}^{\infty} \frac{x^n}{n!} = 1 + x + \frac{x^2}{2!} + \frac{x^3}{3!} + \frac{x^4}{4!} + \cdots.$$
With $x = 1$ we obtain a series for e itself:
$$e^1 = e = \sum_{n=0}^{\infty} \frac{1}{n!} = 1 + 1 + \frac{1}{2!} + \frac{1}{3!} + \frac{1}{4!} + \cdots.^\dagger \qquad (4.36)$$
As with both $\sqrt{2}$ and π, we will assume e is rational and demonstrate that such a supposition leads to a contradiction.

If rational, then by definition e is expressible as the ratio $\frac{a}{b}$ for $a, b \in \mathbb{Z}$ where $b \neq 0$.‡ Exploiting the value b for all it's worth,

---

†For a derivation of this formula using the definition of e see §A.5.
‡Because $e > 0$ (this is obvious from (4.36)), both $a, b > 0$, so our future action of using b as an upper bound in a summation is justified.

§4.6. On the Irrationality of e

notice the sum in (4.36) is, in a larger expansion,

$$e = 1 + 1 + \frac{1}{2!} + \frac{1}{3!} + \cdots + \frac{1}{b!} + \frac{1}{(b+1)!} + \cdots.$$

Hence with the definition

$$R := \sum_{n=b+1}^{\infty} \frac{1}{n!} = \frac{1}{(b+1)!} + \frac{1}{(b+2)!} + \cdots$$

we have that

$$e = \left(\sum_{n=0}^{b} \frac{1}{n!}\right) + R \implies R = e - \sum_{n=0}^{b} \frac{1}{n!}.$$

We now define a new value λ by

$$\lambda := b! \times R = b! \left(e - \sum_{n=0}^{b} \frac{1}{n!}\right). \qquad (4.37)$$

Using our assumption $e = \frac{a}{b}$,

$$\lambda = b! \left(\frac{a}{b} - \sum_{n=0}^{b} \frac{1}{n!}\right) = a(b-1)! - \sum_{n=0}^{b} \frac{b!}{n!}.$$

Clearly the value $a(b-1)! \in \mathbb{Z}$ as both $a, b \in \mathbb{Z}$. For the sum

$$\sum_{n=0}^{b} \frac{b!}{n!} = b! + b! + \frac{b!}{2!} + \frac{b!}{3!} + \cdots + \frac{b!}{b!}$$

each term is clearly an integer, thus the total sum must also be an integer. Altogether we have shown the following important result:

> λ is an integer.

The punch line of the proof is near. What lies ahead is us proving that $\lambda > 0$ and that $\lambda < 1$. This last inequality formulates the contradiction: λ cannot both be an integer and be between zero and one because there are no integers in this range.

To prove $\lambda > 0$ we recall the definition in (4.37). From (4.36), (4.37) is equivalent to

$$\lambda = b! \left(\sum_{n=0}^{\infty} \frac{1}{n!} - \sum_{n=0}^{b} \frac{1}{n!} \right) = \sum_{n=b+1}^{\infty} \frac{b!}{n!}.$$

Each term in this series is strictly positive, therefore $\lambda > 0$.

The proof of $\lambda < 1$ requires a bit of creativity. To do this, we'll examine the fraction $\frac{b!}{n!}$ in the sum above. Notice for $n \geq b+1$, the ratio

$$\frac{b!}{n!} = \frac{b \times (b-1) \times (b-2) \times \cdots \times 2 \times 1}{n \times (n-1) \times \cdots \times b \times (b-1) \times \cdots \times 2 \times 1}$$

$$= \frac{1}{n \times (n-1) \times \cdots \times (b+2) \times (b+1)}.$$

This expansion facilitates our assurance of the following inequality:

$$\underbrace{\frac{1}{n \times (n-1) \times \cdots \times (b+2) \times (b+1)}}_{n-b \text{ factors}} < \frac{1}{(b+1)^{n-b}}.$$

That is,

$$\frac{b!}{n!} < \frac{1}{(b+1)^{n-b}}. \tag{4.38}$$

To reiterate, this inequality is true for all n such that $n \geq b+1$. Now the sum

$$\sum_{n=b+1}^{\infty} \frac{b!}{n!}$$

is over n such that $n \geq b+1$, so the inequality in (4.38) applies to every term. This guarantees the inequality

$$\lambda = \sum_{n=b+1}^{\infty} \frac{b!}{n!} < \sum_{n=b+1}^{\infty} \frac{1}{(b+1)^{n-b}}.$$

If we now change the index of summation $n = b+1$ to $k = 1$ by defining $k = n - b$, the equation above becomes

$$\lambda = \sum_{n=b+1}^{\infty} \frac{b!}{n!} < \sum_{k=1}^{\infty} \frac{1}{(b+1)^k}. \tag{4.39}$$

The sum on the right should look familiar—it is an infinite geometric series with $\omega = \frac{1}{b+1}$. Using the formula

$$\sum_{n=1}^{\infty} C\omega^n = \frac{C}{1-\omega},$$

which is true provided $|\omega| < 1$ (derived in §A.2), it follows that

$$\sum_{k=1}^{\infty} \frac{1}{(b+1)^k} = \frac{\frac{1}{b+1}}{1 - \frac{1}{b+1}} = \frac{1}{b} < 1$$

because $b > 1$.[†] By (4.39) and our proof that $\lambda > 0$, we have shown

$$0 < \lambda < \frac{1}{b} < 1. \tag{4.40}$$

But λ is an integer, so (4.40) is absurd because no integers exist between zero and one. Thus we have reached a contradiction, rendering our assumption that e is rational incorrect. Consequently, Euler's number e is irrational.

§4.7 THE BASEL PROBLEM

We've encountered many famous results in mathematics—the most notable being Euler's identity $e^{i\pi} + 1 = 0$. This section presents yet another famous result—one whose origin heralds a revolution in the field of mathematical analysis. We speak, of course, of the coveted *Basel problem*, which asks to determine the value to which the following series converges:

$$\sum_{n=1}^{\infty} \frac{1}{n^2} = \frac{1}{1^2} + \frac{1}{2^2} + \frac{1}{3^2} + \frac{1}{4^2} + \frac{1}{5^2} + \cdots. \tag{4.41}$$

We refer to (4.41) as either the *Basel sum* or *Basel series*.

While there is no mathematician named "Basel," the problem retrieves its name from Basel, Switzerland—the hometown of the

---

[†] $b \neq 1$ because otherwise $e = \frac{a}{b} = a$ saying that e is an integer. This is certainly not the case because e has a decimal expansion.

distinguished Leonhard Euler. It should come as no surprise, then, that the Basel problem was first cracked by Euler. In this section, we follow his original procedure for solving the Basel problem—an approach that deduces the limit of the Basel sum via a mere coincidence.

We'll begin with a short discussion on a particular type of series called *p-series*. These are any infinite series with the structure

$$\sum_{n=1}^{\infty} \frac{1}{n^p} \qquad (4.42)$$

where $p \in \mathbb{R}$. Using a convergence test called an *integral test*, it is easy to show that for $p > 1$, (4.42) converges while for $p \leq 1$, (4.42) diverges.† We see immediately that the Basel problem is to find the limit of the p-series with $p = 2$. Since $p > 1$, (4.41) converges. Evidently, though, this tells us nothing about the limit of the series. As you'll intuit, knowing whether a series converges or diverges is an interesting and often helpful piece of information. But, for convergent series, what is infinitely more interesting is knowing *to what value* the series converges. And this is precisely the motivation behind the Basel problem.

To begin, recall that any polynomial $P(x)$ can be expressed as a product of its *linear factors* derived using the roots of $P(x)$. For instance, the polynomial $P(x) := x^2 - x - 6$, which has roots $\{3, -2\}$, can be reformulated as $P(x) = (x-3)(x+2)$. Here each factor $(x-3)$ and $(x+2)$ is a linear factor of $P(x)$ (this is mere factorization). We can apply this idea to the more general polynomial $R(x)$ defined by

$$R(x) := a_n x^n + a_{n-1} x^{n-1} + \cdots + a_1 x + a_0.$$

---

†Note for $p = 1$, (4.42) becomes the divergent harmonic series (§4.3), in agreement with the stated bounds for the convergence/divergence of (4.42). We expand on this relationship in §4.11.

§4.7. The Basel Problem

This equation will have n roots, which we organize in the set $\mathcal{R} := \{\alpha_1, \alpha_2, \cdots, \alpha_n\}$. Reformulating $R(x)$ as a product of its linear factors yields

$$R(x) = a_n(x - \alpha_1)(x - \alpha_2) \cdots (x - \alpha_n),$$

where all $\alpha_i \in \mathcal{R}$. Notice the same result holds even after dividing $R(x)$ by the negative of all its roots:

$$R(x) = C\left(1 - \frac{x}{\alpha_1}\right)\left(1 - \frac{x}{\alpha_2}\right) \cdots \left(1 - \frac{x}{\alpha_n}\right), \quad (4.43)$$

where $C = a_n(-\alpha_1)(-\alpha_2) \cdots (-\alpha_n)$ is a constant.

This process is predictably familiar to you. We bring this up because the same idea can be applied to Maclaurin polynomials. Rather than listing them in an infinite series, it is possible to separate them into a product of its linear factors, similar in form to (4.43). We demonstrate this with the sine function.

Recall from §4.1 that $\sin(x)$ is specified by the Maclaurin series

$$\sin(x) = \sum_{n=0}^{\infty} \frac{(-1^n)x^{2n+1}}{(2n+1)!} = x - \frac{x^3}{3!} + \frac{x^5}{5!} - \frac{x^7}{7!} + \cdots. \quad (4.44)$$

Moreover, in §2.4 we noted that $\sin(x)$ will equal zero whenever x is an integer multiple of π—that is, whenever $x = m\pi$ for $m \in \mathbb{Z}$. Therefore, the roots of $\sin(x)$ are contained in the (infinite) set $\{\cdots, -3\pi, -2\pi, -\pi, 0, \pi, 2\pi, 3\pi, \cdots\}$. It follows from expanding (4.44) as a product of linear factors that

$$\sin(x) = Ax(x-\pi)(x+\pi)(x-2\pi)(x+2\pi)(x-3\pi)(x+3\pi)\cdots, \quad (4.45)$$

where A is some constant. As we did with (4.43), we can divide (4.45) by each of its roots (with the exception of zero) to express (4.45) as

$$\sin(x) = Bx\left(1 - \frac{x}{\pi}\right)\left(1 + \frac{x}{\pi}\right)\left(1 - \frac{x}{2\pi}\right)\left(1 + \frac{x}{2\pi}\right) \cdots \quad (4.46)$$

where B is yet another constant.

To determine B, we'll consider (4.46) divided by the first factor x:

$$\frac{\sin(x)}{x} = B\left(1 - \frac{x}{\pi}\right)\left(1 + \frac{x}{\pi}\right)\left(1 - \frac{x}{2\pi}\right)\left(1 + \frac{x}{2\pi}\right)\cdots.$$

Taking the limit as $x \to 0$, we obtain

$$\lim_{x \to 0} \frac{\sin(x)}{x} = \lim_{x \to 0} B\left(1 - \frac{x}{\pi}\right)\left(1 + \frac{x}{\pi}\right)\left(1 - \frac{x}{2\pi}\right)\left(1 + \frac{x}{2\pi}\right)\cdots$$
$$= B\,(1)\,(1)\,(1)\,(1)\cdots$$
$$= B.$$

But we proved in §2.3 that the limit

$$\lim_{x \to 0} \frac{\sin(x)}{x} = 1,$$

forcing the conclusion that $B = 1$ and, additionally, that (4.46) is just

$$\sin(x) = x\left(1 - \frac{x}{\pi}\right)\left(1 + \frac{x}{\pi}\right)\left(1 - \frac{x}{2\pi}\right)\left(1 + \frac{x}{2\pi}\right)\cdots.$$

To simplify this expression, observe that each nonzero integer k participates in a factor-pair of the form

$$\left(1 - \frac{x}{k\pi}\right)\left(1 + \frac{x}{k\pi}\right) = \left(1 - \frac{x^2}{k^2\pi^2}\right).$$

Hence,

$$\sin(x) = x\left(1 - \frac{x^2}{\pi^2}\right)\left(1 - \frac{x^2}{4\pi^2}\right)\left(1 - \frac{x^2}{9\pi^2}\right)\cdots.$$

In product notation:

$$\sin(x) = x \prod_{k=1}^{\infty}\left(1 - \frac{x^2}{k^2\pi^2}\right). \qquad (4.47)$$

This is a truly remarkable formula. In this book, not only have we formulated a series expansion for the sine function, but also now a

§4.7. The Basel Problem

product expansion. In the following section we will make good use of (4.47) and squeeze out a fascinating formula for π.

In the mean time, let's expand the product in (4.47). Multiplying the first two factors, we obtain

$$\sin(x) = \left(x - \frac{x^3}{\pi^2}\right)\left(1 - \frac{x^2}{4\pi^2}\right)\left(1 - \frac{x^2}{9\pi^2}\right)\left(1 - \frac{x^2}{16\pi^2}\right)\cdots.$$

Similarly with the next (third) factor:

$$\sin(x) = \left(x - \frac{x^3}{4\pi^2} - \frac{x^3}{\pi^2} + \frac{x^5}{4\pi^4}\right)\left(1 - \frac{x^2}{9\pi^2}\right)\left(1 - \frac{x^2}{16\pi^2}\right)\cdots$$

$$= \left[x - \frac{x^3}{\pi^2}\left(1 + \frac{1}{4}\right) + \frac{x^5}{4\pi^4}\right]\left(1 - \frac{x^2}{9\pi^2}\right)\left(1 - \frac{x^2}{16\pi^2}\right)\cdots$$

And once more with the next (fourth) factor:

$$\sin(x) = \left[x - \frac{x^3}{9\pi^2} - \frac{x^3}{\pi^2}\left(1 + \frac{1}{4}\right)\right.$$

$$\left. + \left(\frac{x^5}{4\pi^4} + \frac{x^5}{9\pi^4} + \frac{x^5}{36\pi^4}\right) - \frac{x^7}{36\pi^6}\right]\left(1 - \frac{x^2}{16\pi^2}\right)\cdots$$

$$= \left[x - \frac{x^3}{\pi^2}\left(1 + \frac{1}{4} + \frac{1}{9}\right) + \frac{x^5}{\pi^4}\left(\frac{1}{4} + \frac{1}{9} + \frac{1}{36}\right)\right.$$

$$\left. - \frac{x^7}{36\pi^6}\right]\left(1 - \frac{x^2}{16\pi^2}\right)\cdots. \qquad (4.48)$$

Before this product gets out of control, notice that the second term in the polynomial expansion incorporates the Basel sum. More explicitly, the second term above reads

$$-\frac{x^3}{\pi^2}\underbrace{\left(1 + \frac{1}{4} + \frac{1}{9} + \frac{1}{16} + \cdots\right)}_{\text{the Basel sum}}$$

$$= -\frac{x^3}{\pi^2}\left(\frac{1}{1^2} + \frac{1}{2^2} + \frac{1}{3^2} + \frac{1}{4^2} + \cdots\right).$$

We are now almost there. Because (4.48) is the sine function, it must equal the Maclaurin expansion of $\sin(x)$. Symbolically, this means that

$$\sin(x) = x - \frac{x^3}{3!} + \frac{x^5}{5!} - \cdots$$

$$= x - \frac{x^3}{\pi^2}\left(\frac{1}{1^2} + \frac{1}{2^2} + \frac{1}{3^2} + \frac{1}{4^2} + \cdots\right) + \cdots.$$

But the one-to-one correspondence between the term and degree of the variable x in each series requires

$$-\frac{x^3}{3!} = -\frac{x^3}{\pi^2}\left(\frac{1}{1^2} + \frac{1}{2^2} + \frac{1}{3^2} + \frac{1}{4^2} + \cdots\right).$$

Rearranging, we stumble upon the astonishing result

$$\sum_{n=1}^{\infty} \frac{1}{n^2} = \frac{1}{1^2} + \frac{1}{2^2} + \frac{1}{3^2} + \frac{1}{4^2} + \cdots = \frac{\pi^2}{6}. \qquad (4.49)$$

We stress that this is purely a coincidence—(4.49) was obtained by pure manipulation of the Maclaurin series for $\sin(x)$, no clever mathematics was really needed. Amazing, right?

Before closing the section, we note a splendid corollary of (4.49). Consider grouping the "even" and "odd" terms in the Basel series. By this we mean writing (4.49) as

$$\frac{1}{1^2} + \frac{1}{2^2} + \frac{1}{3^2} + \cdots = \overbrace{\left(\frac{1}{2^2} + \frac{1}{4^2} + \cdots\right)}^{\text{"even" terms}} + \overbrace{\left(\frac{1}{1^2} + \frac{1}{3^2} + \cdots\right)}^{\text{"odd" terms}}.$$

In summation notation, the even sum is

$$\sum_{n=1}^{\infty} \frac{1}{(2n)^2} = \frac{1}{2^2} + \frac{1}{4^2} + \frac{1}{6^2} + \cdots \qquad (4.50)$$

while the odd sum is

$$\sum_{n=1}^{\infty} \frac{1}{(2n-1)^2} = \frac{1}{1^2} + \frac{1}{3^2} + \frac{1}{5^2} + \cdots.$$

Together, these form the Basel series

$$\frac{\pi^2}{6} = \sum_{n=1}^{\infty} \frac{1}{(2n)^2} + \sum_{n=1}^{\infty} \frac{1}{(2n-1)^2}. \qquad (4.51)$$

§4.8. The Wallis Product

Observe that (4.50) is simply

$$\sum_{n=1}^{\infty} \frac{1}{(2n)^2} = \frac{1}{4}\left(\sum_{n=1}^{\infty} \frac{1}{n^2}\right) = \frac{\pi^2}{24},$$

which is a spectacular result in itself. But from (4.51), yet another result follows:

$$\frac{\pi^2}{6} = \frac{\pi^2}{24} + \sum_{n=1}^{\infty} \frac{1}{(2n-1)^2} \implies \sum_{n=1}^{\infty} \frac{1}{(2n-1)^2} = \frac{\pi^2}{8}.$$

To us, these series are eye-candy. We hope they are for you too.

§4.8 The Wallis Product

The solution to the Basel problem is quite an unexpected result. While it is difficult to top, we try to do so with the product to which this section is devoted. Named after the seventeenth century mathematician John Wallis, the *Wallis product* concerns the infinite product

$$\prod_{k=1}^{\infty} \frac{(2k)(2k)}{(2k-1)(2k+1)} = \frac{2}{1} \times \frac{2}{3} \times \frac{4}{3} \times \frac{4}{5} \times \frac{6}{5} \times \frac{6}{7} \times \cdots. \qquad (4.52)$$

Evidently, the problem is to figure out to what value (4.52) converges. We do this using two separate methods. The first utilizes a delightful integral within which the Wallis product is curiously embedded, while the second stems from the product representation for the sine function derived in the previous section.

We begin with the integral

$$a_k = \int_0^{\frac{\pi}{2}} \sin^k(x)\,dx, \qquad (4.53)$$

where $k \in \mathbb{N}_0$. Notice that in using the notation a_k for (4.53), we can express the integral

$$\int_0^{\frac{\pi}{2}} \sin^{k-1}(x)\,dx$$

as a_{k-1}. Importantly, the integral given by a_0 would imply $k = 0$ so that

$$a_0 = \int_0^{\frac{\pi}{2}} \sin^0(x)dx = \int_0^{\frac{\pi}{2}} dx = \frac{\pi}{2}.$$

Similarly, that for a_1, which implies $k = 1$, would yield

$$a_1 = \int_0^{\frac{\pi}{2}} \sin^1(x)dx = \int_0^{\frac{\pi}{2}} \sin(x)dx = -\cos(x)\Big|_0^{\frac{\pi}{2}} = 1.$$

Remember these results for a_0 and a_1 as they are vital to the analysis that follows.

As you've probably guessed, to obtain the Wallis product we need to evaluate (4.53) for all values of k—and what better way to do so than with integration by parts. Let

$$u = \sin^{k-1}(x) \implies du = (k-1)\sin^{k-2}(x)\cos(x)dx$$
$$dv = \sin(x)dx \implies v = -\cos(x)$$

so that

$$a_k = \int_0^{\frac{\pi}{2}} u\,dv = uv\Big|_0^{\frac{\pi}{2}} - \int_0^{\frac{\pi}{2}} v\,du.$$

Back substituting, we obtain

$$a_k = \underbrace{-\sin^{k-1}(x)\cos(x)\Big|_0^{\frac{\pi}{2}}}_{\text{this term vanishes}} + (k-1)\int_0^{\frac{\pi}{2}} \sin^{k-2}(x)\cos^2(x)dx.$$

Hence, we have

$$a_k = (k-1)\int_0^{\frac{\pi}{2}} \sin^{k-2}(x)\cos^2(x)dx. \tag{4.54}$$

Using the identity $\cos^2(x) = 1 - \sin^2(x)$, it follows that (4.54) is equivalent to

$$a_k = (k-1)\underbrace{\int_0^{\frac{\pi}{2}} \sin^{k-2}(x)dx}_{\text{this is } a_{k-2}} - (k-1)\underbrace{\int_0^{\frac{\pi}{2}} \sin^k(x)dx}_{\text{this is } a_k}.$$

§4.8. THE WALLIS PRODUCT

More compactly,
$$a_k = (k-1)a_{k-2} - (k-1)a_k.$$

Solving for a_k by itself yields
$$a_k = \frac{k-1}{k}a_{k-2}. \tag{4.55}$$

It will be useful to express (4.55) with two separate values for k—namely, $k_1 = 2n$ and $k_2 = 2n+1$, for $n \in \mathbb{N}_0$. This way we account for all even and odd values the subscript k in a_k can take on. Doing this small transformation generates the two recurrence relationships

$$a_{k_1} = a_{2n} = \frac{2n-1}{2n}a_{2n-2} \tag{4.56}$$

$$a_{k_2} = a_{2n+1} = \frac{2n}{2n+1}a_{2n-1}. \tag{4.57}$$

Recall from §2.5 that in order to solve a recurrence relationship like that in (4.56), we first need to compute a_{2n-2}; but in order to compute a_{2n-2}, we first need to compute a_{2n-4}; but in order to compute a_{2n-4}, etc. This process trickles down to the smallest value of n allowed by the domain. Since we have defined $n \in \mathbb{N}_0$, it follows the minimum value of n is $n = 0$. Piecing this together with the recurrence relation in (4.56), it follows that for $n = 1$,

$$a_{2(1)} = a_2 = \frac{2(1)-1}{2(1)}a_{2(1)-2} = \frac{1}{2}a_0 = \frac{1}{2} \times \frac{\pi}{2}$$

because, as we derived above, $a_0 = \frac{\pi}{2}$. Setting $n = 2$, we find from (4.56) that

$$a_{2(2)} = a_4 = \frac{2(2)-1}{2(2)}a_2 = \frac{3}{4} \times \frac{1}{2} \times \frac{\pi}{2}$$

because $a_2 = \frac{1}{2} \times \frac{\pi}{2}$. Continuing this process for arbitrary n, it follows that

$$a_{2n} = \frac{\pi}{2} \times \frac{1}{2} \times \frac{3}{4} \times \frac{5}{6} \times \cdots \times \frac{2n-1}{2n}. \tag{4.58}$$

If we do the same for the recurrence relation in (4.57)—starting at $n = 1$, then $n = 2$, and so forth—we establish that

$$a_{2n+1} = 1 \times \frac{2}{3} \times \frac{4}{5} \times \frac{6}{7} \times \frac{8}{9} \times \cdots \times \frac{2n}{2n+1}.^\dagger \qquad (4.59)$$

Examining (4.58) and (4.59), we can see pieces of the Wallis product sewn in. To reveal the product explicitly, we must first analyze how the sine function (the integrand of the integral defined by a_{2n}) relates to itself under different inputs for n. In other words, we must note that for all $n \in \mathbb{N}_0$,

$$\sin^{2n+1}(x) \leq \sin^{2n}(x) \leq \sin^{2n-1}(x). \qquad (4.60)$$

Under the interval $0 \leq x \leq \frac{\pi}{2}$, the sine function, regardless of its power, is positive. Thus, it follows from (4.60) that

$$\int_0^{\frac{\pi}{2}} \sin^{2n+1}(x)\,dx \leq \int_0^{\frac{\pi}{2}} \sin^{2n}(x)\,dx \leq \int_0^{\frac{\pi}{2}} \sin^{2n-1}(x)\,dx.$$

That is,

$$a_{2n+1} \leq a_{2n} \leq a_{2n-1} \implies 1 \leq \frac{a_{2n}}{a_{2n+1}} \leq \frac{a_{2n-1}}{a_{2n+1}}. \qquad (4.61)$$

But we know from (4.57) that $a_{2n+1} = \frac{2n}{2n-1} a_{2n-1}$. This allows the fraction on the right-hand side of (4.61) to be reformulated as

$$\frac{a_{2n-1}}{a_{2n+1}} = \frac{a_{2n-1}}{\frac{2n}{2n-1}(a_{2n-1})} = \frac{2n-1}{2n}.$$

Therefore, the compound inequality in (4.61) is equivalent to

$$1 \leq \frac{a_{2n}}{a_{2n+1}} \leq \frac{2n-1}{2n}. \qquad (4.62)$$

Because the Wallis product is an *infinite* product, we require some notion of infinite arithmetic in our final answer. Letting $n \to \infty$ in (4.62) will do the trick. In this limit, we obtain

$$\lim_{n \to \infty} 1 \leq \lim_{n \to \infty} \frac{a_{2n}}{a_{2n+1}} \leq \lim_{n \to \infty} \frac{2n-1}{2n}.$$

---

†The reader is encouraged to derive this expression. *Hint:* you require the result $a_1 = 1$, which we derived previously.

§4.8. THE WALLIS PRODUCT

Here, as $n \to \infty$ the fraction $\frac{2n-1}{2n} \to 1$ and, obviously, $1 \to 1$. Hence,
$$1 \leq \lim_{n \to \infty} \frac{a_{2n}}{a_{2n+1}} \leq 1.$$
Similar to the limit we encountered in §2.3, the central expression is squeezed between one and one,[†] forcing the conclusion that
$$\lim_{n \to \infty} \frac{a_{2n}}{a_{2n+1}} = 1. \tag{4.63}$$
But from (4.58) and (4.59) we know what a_{2n} and a_{2n+1} are. Upon substituting these equations into (4.63), we arrive at the limit
$$\lim_{n \to \infty} \frac{a_{2n}}{a_{2n+1}} = \frac{\frac{\pi}{2} \times \frac{1}{2} \times \frac{3}{4} \times \cdots}{1 \times \frac{2}{3} \times \frac{4}{5} \times \cdots} = 1$$
$$\implies \frac{\pi}{2} \times \frac{1}{2} \times \frac{3}{4} \times \cdots = 1 \times \frac{2}{3} \times \frac{4}{5} \times \cdots.$$
After a little algebraic manipulation, the Wallis product comes right out:
$$\prod_{k=1}^{\infty} \frac{(2k)(2k)}{(2k-1)(2k+1)} = \frac{2}{1} \times \frac{2}{3} \times \frac{4}{3} \times \frac{4}{5} \times \frac{6}{5} \times \frac{6}{7} \times \cdots = \frac{\pi}{2}.$$
We mentioned at the start of this section that two methods would be used to compute the limit of the Wallis product. For the second (much shorter) method, we utilize the sine function's infinite product derived in §4.7 by factoring its Maclaurin series:
$$\sin(x) = x \prod_{k=1}^{\infty} \left(1 - \frac{x^2}{k^2 \pi^2}\right) = x\left(1 - \frac{x^2}{\pi^2}\right)\left(1 - \frac{x^2}{4\pi^2}\right)\left(1 - \frac{x^2}{9\pi^2}\right) \cdots. \tag{4.64}$$
Writing (4.64) with input $x = \frac{\pi}{2}$ we obtain
$$\sin\left(\frac{\pi}{2}\right) = \frac{\pi}{2} \prod_{k=1}^{\infty} \left(1 - \frac{1}{4k^2}\right) = \frac{\pi}{2} \prod_{k=1}^{\infty} \left(\frac{4k^2 - 1}{4k^2}\right) = 1.$$

[†]Mimicking the footnote on page 38, here the word "squeezed" is used intentionally since (4.63) follows from something called the *squeeze theorem*. See the referenced footnote for additional information.

Note that the numerator $4k^2 - 1 = (2k-1)(2k+1)$ and the denominator $4k^2 = (2k)(2k)$. Therefore,

$$1 = \frac{\pi}{2} \prod_{k=1}^{\infty} \frac{(2k-1)(2k+1)}{(2k)(2k)}.$$

Transferring the product to the left-side of the equation, the Wallis product follows immediately:

$$\prod_{k=1}^{\infty} \frac{(2k)(2k)}{(2k-1)(2k+1)} = \frac{\pi}{2}. \qquad (4.65)$$

This second method is naturally a more beautiful derivation. Trudging though the calculus with the first method, however, makes the result even more satisfying.

At the end of §3.7, we introduced the double factorial function

$$m!! := \begin{cases} m \times (m-2) \times (m-4) \times \cdots \times 3 \times 1 & \text{if } m \text{ is odd} \\ m \times (m-2) \times (m-4) \times \cdots \times 2 \times 1 & \text{if } m \text{ is even.} \end{cases}$$

Using these expression, we can write the Wallis product as a traditional limit, without the need for product notation. The factors of $2k$ in the numerator of (4.65) correspond to $(2k)!!$ in the infinite product. Similarly, the factors $(2k-1)$ and $(2k+1)$ correspond to $(2k-1)!!$ and $(2k+1)!!$, respectively. Together, these formulate the fraction

$$\frac{(2k)!!(2k)!!}{(2k-1)!!(2k+1)!!}.$$

Taking the limit as $k \to \infty$ thus yields the Wallis product:

$$\lim_{k \to \infty} \frac{(2k)!!(2k)!!}{(2k-1)!!(2k+1)!!} = \frac{2}{1} \times \frac{2}{3} \times \frac{4}{3} \times \frac{4}{5} \times \cdots = \frac{\pi}{2}. \qquad (4.66)$$

An intriguing limit, to say the least. While impressing others with the Wallis product is dignifying enough, greater status is obtained by first bewildering others with (4.66), then proving the Wallis product.

§4.9 Riemann's Rearrangement Theorem

Infinity is a convoluted and counterintuitive notion. We have met many enthralling applications of infinity spanning from fractals to products. In this section, we explore the nineteenth century mathematician Bernhard Riemann's well-known *rearrangement theorem*, which concerns a peculiar type of infinite series called *conditionally convergent series*.

We begin our discussion by studying *alternating series*. These are any infinite series whose terms oscillate between positive and negative values. The most elementary example is the sum

$$\sum_{n=0}^{\infty} (-1)^n = 1 - 1 + 1 - 1 + 1 - 1 + \cdots. \qquad (4.67)$$

In more general notation, (4.67) is equivalent to the series

$$\sum_{n=0}^{\infty} a_n$$

provided $a_n = (-1)^n$. This notation is useful because we can now define new series through operations on a_n. For instance, with $a_n = (-1)^n$ the series

$$\sum_{n=0}^{\infty} |a_n| = 1 + 1 + 1 + 1 + 1 + 1 + \cdots.$$

This will be used extensively shortly.

We now consider the series

$$A := \sum_{n=0}^{\infty} a_n = L_1$$

where $L_1 \in \mathbb{R}$ and finite, implying A is convergent. More specifically, A is said to be *absolutely convergent* if, and only if, both A and the series

$$\sum_{n=0}^{\infty} |a_n| \qquad (4.68)$$

converge. Conversely, if A converges but (4.68) diverges, then A is *conditionally convergent*. It is conditionally convergent series for which the rearrangement theorem applies. To get to the heart of this theorem, however, we still require a crucial piece of information, which has to do with the addition of two or more infinite series.

Suppose we have two infinite series

$$A := \sum_{n=0}^{\infty} a_n = L_1 \text{ and } B := \sum_{n=0}^{\infty} b_n = L_2. \qquad (4.69)$$

For now, we neglect to specify whether L_1 and L_2 are finite or infinite, whether A and B are convergent or divergent. Regardless of their limits, it is true that the sum

$$A + B = \sum_{n=0}^{\infty} a_n + \sum_{n=0}^{\infty} b_n$$
$$= \underbrace{(a_0 + a_1 + a_2 + \cdots)}_{L_1} + \underbrace{(b_0 + b_1 + b_2 + \cdots)}_{L_2}.$$

Since both series have an infinite number of terms, we can pair the terms in each series by the order in which they appear:

$$A + B = \sum_{n=0}^{\infty} (a_n + b_n) = (a_0 + b_0) + (a_1 + b_1) + (a_2 + b_2) + \cdots.$$
(4.70)

Clearly, from the definitions of A and B in (4.69), the sum $A + B = L_1 + L_2$. Thus, it follows from (4.70) that

$$A + B = \sum_{n=0}^{\infty} (a_n + b_n) = L_1 + L_2. \qquad (4.71)$$

Observe that if both series A and B are convergent, then the sum $L_1 + L_2$ is necessarily finite, which suggests the series $A + B$ is also convergent. However, if either A or B diverges towards $\pm$ infinity, then, necessarily, the series $A + B$ must also diverge towards $\pm$ infinity. If, however, *both* series A and B are divergent and tend off

§4.9. Riemann's Rearrangement Theorem

towards $\pm$ infinity, then the series $A + B$ is *not* necessarily divergent.

To illustrate this, suppose the series A diverges and the series B converges. Let

$$C := B - A = \sum_{n=0}^{\infty} (b_n - a_n).$$

Because A diverges and B converges, C necessarily diverges. Combining the two *divergent* series A and C through addition yields

$$A + C = \sum_{n=0}^{\infty} a_n + \sum_{n=0}^{\infty} (b_n - a_n) = \sum_{n=0}^{\infty} b_n.$$

But this series is just B, which we know converges. So the sum of two divergent series is not necessarily a divergent one. This fact is vital to the rearrangement theorem, for which we are now ready:

Theorem 4.4 (Riemann's Rearrangement Theorem). *For any conditionally convergent series R there exists a rearrangement in the terms of R such that it converges to whatever value you like.*

At first, Theorem 4.4 is clearly absurd. After all, how can a series converge to two finite limits L_1 and L_2 with $L_1 \neq L_2$? To address this seemingly non-sequitur statement, suppose we have the conditionally convergent series

$$R := \sum_{n=0}^{\infty} (-1)^n r_n = r_0 - r_1 + r_2 - r_3 + \cdots = L$$

where L is finite. Because R is conditionally convergent, the series

$$|R| = \sum_{n=0}^{\infty} |(-1)^n r_n| = \sum_{n=0}^{\infty} r_n = r_0 + r_1 + r_2 + r_3 + \cdots$$

is necessarily divergent. Let's break the series R into its positive and negative parts, giving us

$$R^+ = \sum_{n=0}^{\infty} r_{2n} = r_0 + r_2 + r_4 + \cdots$$

and

$$R^- = (-1) \times \sum_{n=0}^{\infty} r_{2n+1} = (-1) \times (r_1 + r_3 + r_5 + \cdots).$$

What can we say about the convergence/divergence of R^+ and R^-? Well, from the definitions of both R^+ and R^-, we can write

$$R = R^+ + R^- = \sum_{n=0}^{\infty} r_{2n} - \sum_{n=0}^{\infty} r_{2n+1}.$$

Because R converges, we know from (4.71) that the sum of two series converges if both series in the sum converge or, possibly, if both diverge. Because R is conditionally convergent and the series $|R|$ diverges, it follows that

$$|R| = \sum_{n=0}^{\infty} r_{2n} + \sum_{n=0}^{\infty} r_{2n+1}$$

is divergent. Thus, both R^+ and R^- must diverge.[†] In other words, for any conditionally convergent series of the form

$$R = \sum_{n=0}^{\infty} (-1)^n r_n,$$

the positive terms form a divergent series as do the negative terms, even though, when summed together, they converge to a finite limit.

This is a powerful result, as it implies that for any conditionally convergent series there is always an abundance of terms that tend off towards both positive and negative infinity. We can thus redirect a sum in any direction we like by simply rearranging the terms of the sum; and this is exactly the rearrangement theorem. More specifically, if we want to make the conditionally convergent series R converge to, say, 42, we can start by summing a few of

---

[†]Both cannot converge because the sum of two finite values is always finite. Moreover, one cannot converge while the other diverges because then both R and $|R|$ would diverge (why?).

§4.9. Riemann's Rearrangement Theorem

the positive terms so that we overshoot 42, then tack on a few of the negative terms so we undershoot 42, then add positive terms to overshoot, and so forth. Performing this forever, the series would appear to close in on 42, rightfully prompting a mathematician to state the limit of the series is indeed 42. Obviously, 42 need not be the chosen value (even if it is the meaning of the universe). Due to the abundance of positive and negative terms, each with sums diverging towards $\pm$ infinity, we can manipulate the series to obtain any finite value we like, even irrational numbers like e and π.

To provide a more explicit example of Theorem 4.4, we consider the Maclaurin series for the function $\log(x+1)$:

$$\log(x+1) = \sum_{n=1}^{\infty} \frac{(-1)^{n-1} x^n}{n} = x - \frac{x^2}{2} + \frac{x^3}{3} - \frac{x^4}{4} + \cdots .^\dagger$$

Setting $x = 1$, it follows that

$$\log(2) = \sum_{n=1}^{\infty} \frac{(-1)^n}{n} = 1 - \frac{1}{2} + \frac{1}{3} - \frac{1}{4} + \frac{1}{5} - \frac{1}{6} + \cdots . \quad (4.72)$$

We emphasize that the limit of this series is $\log(2) \approx 0.69315$, as determined from the Maclaurin series definition. In considering the absolute value of (4.72), we obtain the harmonic series

$$\sum_{n=1}^{\infty} \frac{1}{n} = 1 + \frac{1}{2} + \frac{1}{3} + \frac{1}{4} + \cdots ,$$

which from §4.3 diverges. Thus, (4.72) is a conditionally convergent series and is therefore susceptible to Riemann's rearrangement theorem. To see how, define

$$S := \sum_{n=1}^{\infty} \frac{(-1)^{n-1}}{n} = 1 - \frac{1}{2} + \frac{1}{3} - \frac{1}{4} + \frac{1}{5} - \frac{1}{6} + \cdots ,$$

---

†This series is know as the *Mercator series* after the seventeenth century mathematician Nicholas Mercator. You are encouraged to derive his sum by using the Maclaurin series definition (4.7) in §4.1.

which we know converges to $\log(2)$. That said, consider the series that would define $2S$, given by

$$2S = \sum_{n=1}^{\infty} \frac{2(-1)^{n-1}}{n} = 2 - 1 + \frac{2}{3} - \frac{1}{2} + \frac{2}{5} - \frac{1}{3} + \frac{2}{7} - \frac{1}{4} + \cdots . \quad (4.73)$$

Similar to S, it can be shown that $2S$ is also conditionally convergent. We will show that a simple rearrangement in the terms of $2S$ makes (4.73) approach $\log(2)$—the limit of S.

To do this, we utilize the following rearrangement: For all terms with an odd denominator and numerator of one, we add it to its counterpart with an equal denominator but with a numerator of two. Combining and rearranging these terms as in

$$2S = (2-1) - \frac{1}{2} + \left(\frac{2}{3} - \frac{1}{3}\right) - \frac{1}{4} + \left(\frac{2}{5} - \frac{1}{5}\right) - \frac{1}{6} + \cdots$$

simplifies the sum to

$$2S = 1 - \frac{1}{2} + \frac{1}{3} - \frac{1}{4} + \frac{1}{5} - \frac{1}{6} + \cdots .$$

But this is just the series S, so we've apparently shown that $2S$ converges to $\log(2)$ and, moreover, that $2S = S$. Now this has to be wrong. The last time we checked $2 \neq 1$, and yet we've apparently just shown the contrary. But where is the error?

The problem lies in our inability to sum an infinite number of terms. Though it appears that the series given by (4.73) is identical to (4.72), upon reaching infinity (an impossible task) the series would indeed equal $2S$. Yet, being conditionally convergent, we can make $2S$ appear to approach S for as long and as many terms as we choose thanks to Theorem 4.4. The moment we stop (we must stop eventually), the remaining terms in (4.73) take over and persuade the instantaneous partial sum to approach its definite limit of $2\log(2) = \log(4)$. It seems, rather unfortunately, the notion of infinity prevents us from concluding that all conditionally convergent

§4.10 VIÈTE'S FORMULA FOR π

So far we have derived a multitude of fanciful formulas for π. In this section, the last of such formulas is derived (at least in this book)—the one we herald as being the most extraordinary.

Our journey begins (somewhat arbitrarily) with the trigonometric addition formulas for sine and cosine:

$$\sin(\alpha + \beta) = \sin(\alpha)\cos(\beta) + \cos(\alpha)\sin(\beta) \tag{4.74}$$

$$\cos(\alpha + \beta) = \cos(\alpha)\cos(\beta) - \sin(\alpha)\sin(\beta). \tag{4.75}$$

We have encountered these many times throughout the book, yet have, up to now at least, neglected their proof: Using Euler's formula, observe

$$\begin{aligned} e^{i(\alpha+\beta)} &= e^{i\alpha} e^{i\beta} \\ &= [\cos(\alpha) + i\sin(\alpha)][\cos(\beta) + i\sin(\beta)] \\ &= [\cos(\alpha)\cos(\beta) - \sin(\alpha)\sin(\beta)] \\ &\quad + i[\sin(\alpha)\cos(\beta) + \cos(\alpha)\sin(\beta)]. \end{aligned}$$

Comparing real and imaginary parts, both (4.74) and (4.75) follow immediately.

Our interest in (4.74) stems from the fact that for $\alpha = \beta = \frac{x}{2}$,

$$\sin\left(\frac{x}{2} + \frac{x}{2}\right) = \sin(x) = 2\sin\left(\frac{x}{2}\right)\cos\left(\frac{x}{2}\right), \tag{4.76}$$

while for (4.75),

$$\cos\left(\frac{x}{2} + \frac{x}{2}\right) = \cos(x) = \cos^2\left(\frac{x}{2}\right) - \sin^2\left(\frac{x}{2}\right). \tag{4.77}$$

For (4.77) we can use $\sin^2\left(\frac{x}{2}\right) = 1 - \cos^2\left(\frac{x}{2}\right)$ to obtain

$$\cos\left(\frac{x}{2}\right) = \sqrt{\frac{1+\cos(x)}{2}}.^\dagger \qquad (4.78)$$

Whether these equations are familiar or not, they do not seem to be closing in on a product for π. To squeeze out π requires an astute observation: (4.76) is a recursive formula. In other words, if we know

$$\sin(x) = 2\sin\left(\frac{x}{2}\right)\cos\left(\frac{x}{2}\right),$$

then we know

$$\sin\left(\frac{x}{2}\right) = 2\sin\left(\frac{x}{2^2}\right)\cos\left(\frac{x}{2^2}\right).$$

Therefore,

$$\sin(x) = 2^2 \sin\left(\frac{x}{2^2}\right)\cos\left(\frac{x}{2}\right)\cos\left(\frac{x}{2^2}\right).$$

Another iteration yields

$$\sin(x) = 2^3 \sin\left(\frac{x}{2^3}\right)\cos\left(\frac{x}{2}\right)\cos\left(\frac{x}{2^2}\right)\cos\left(\frac{x}{2^3}\right).$$

In general, we have shown

$$\sin(x) = 2^n \sin\left(\frac{x}{2^n}\right)\cos\left(\frac{x}{2}\right)\cos\left(\frac{x}{2^2}\right)\cdots\cos\left(\frac{x}{2^n}\right) \qquad (4.79)$$

for arbitrary $n \in \mathbb{N}$. To omit all dependencies on n, we let $n \to \infty$. This generates the product

$$\lim_{n\to\infty} 2^n \sin\left(\frac{x}{2^n}\right) \prod_{k=1}^{n} \cos\left(\frac{x}{2^k}\right). \qquad (4.80)$$

To get a better handle on this limit, we utilize the Maclaurin expansion of $\sin\left(\frac{x}{2^n}\right)$:

$$\sin\left(\frac{x}{2^n}\right) = \sum_{k=0}^{\infty} \frac{(-1)^k \left(\frac{x}{2^n}\right)^{2k+1}}{(2k+1)!} = \sum_{k=0}^{\infty} \frac{(-1)^k x^{2k+1}}{2^{2kn+n}(2k+1)!}.$$

---

†We take the positive solution following the square root because $\cos(0) = 1 \neq -1$.

§4.10. Viète's Formula for π

It follows that the limit in (4.80) is identical to the limit

$$\lim_{n\to\infty} 2^n \left(\sum_{k=0}^{\infty} \frac{(-1)^k x^{2k+1}}{2^{2kn+n}(2k+1)!} \right) \left(\prod_{k=1}^{n} \cos\left(\frac{x}{2^k}\right) \right).$$

For the sum on the left, observe

$$2^n \sum_{k=0}^{\infty} \frac{(-1)^k x^{2k+1}}{2^{2kn+n}(2k+1)!} = \sum_{k=0}^{\infty} \frac{(-1)^k x^{2k+1}}{2^{2kn}(2k+1)!}.$$

Hence, the limit

$$\lim_{n\to\infty} 2^n \sum_{k=0}^{\infty} \frac{(-1)^k x^{2k+1}}{2^{2kn+n}(2k+1)!} = \lim_{n\to\infty} \sum_{k=0}^{\infty} \frac{(-1)^k x^{2k+1}}{2^{2kn}(2k+1)!}$$

$$= \lim_{n\to\infty} x - \frac{x^3}{2^{2n}3!}^{\,0} + \frac{x^5}{2^{4n}5!}^{\,0} - \cdots$$

$$= x.$$

Consequently,

$$\lim_{n\to\infty} 2^n \sin\left(\frac{x}{2^n}\right) \prod_{k=1}^{n} \cos\left(\frac{x}{2^k}\right) = x \prod_{k=1}^{\infty} \cos\left(\frac{x}{2^k}\right).$$

Combining this with (4.79), we obtain yet another incredible formula for the sine function:

$$\sin(x) = x \prod_{k=1}^{\infty} \cos\left(\frac{x}{2^k}\right) \qquad (4.81)$$

$$= x \cos\left(\frac{x}{2}\right) \cos\left(\frac{x}{2^2}\right) \cos\left(\frac{x}{2^3}\right) \cdots.$$

Earlier, we neglected to address the recursive nature of (4.78)— the cosine half-angle formula

$$\cos\left(\frac{x}{2}\right) = \sqrt{\frac{1+\cos(x)}{2}}.$$

From this equation, we have that

$$\cos\left(\frac{x}{2^2}\right) = \sqrt{\frac{1+\cos\left(\frac{x}{2}\right)}{2}}.$$

But $\cos(\frac{x}{2})$ is just (4.78). Therefore,

$$\cos\left(\frac{x}{2^2}\right) = \sqrt{\frac{1+\sqrt{\frac{1+\cos(x)}{2}}}{2}} = \sqrt{\frac{1}{2}+\frac{1}{2}\sqrt{\frac{1}{2}+\frac{\cos(x)}{2}}}.$$

Similarly,

$$\cos\left(\frac{x}{2^3}\right) = \sqrt{\frac{1+\cos\left(\frac{x}{4}\right)}{2}}.$$

Thus, it follows from above that

$$\cos\left(\frac{x}{2^3}\right) = \sqrt{\frac{1}{2}+\frac{1}{2}\sqrt{\frac{1}{2}+\frac{1}{2}\sqrt{\frac{1}{2}+\frac{\cos(x)}{2}}}}.$$

Excitingly, this pattern is true in general:

$$\cos\left(\frac{x}{2^n}\right) = \underbrace{\sqrt{\frac{1}{2}+\frac{1}{2}\sqrt{\frac{1}{2}+\frac{1}{2}\sqrt{\frac{1}{2}+\cdots+\frac{1}{2}\sqrt{\frac{1}{2}+\frac{\cos(x)}{2}}}}}}_{n \text{ square roots}}.$$

(4.82)

This expression is called an *iterated* or *nested radical* because of the repetitive square rooting. We will encounter such an algebraic structure again in §5.1. For now, we use it to finish off our π creation.

Combining (4.82) with the product in (4.81), it follows that

$$\sin(x) = x \times \cos\left(\frac{x}{2}\right) \times \cos\left(\frac{x}{2^2}\right) \times \cos\left(\frac{x}{2^3}\right) \times \cdots$$

$$= x \times \sqrt{\frac{1}{2}+\frac{\cos(x)}{2}} \times \sqrt{\frac{1}{2}+\frac{1}{2}\sqrt{\frac{1}{2}+\frac{\cos(x)}{2}}} \times \cdots.$$

Finally, setting $x = \frac{\pi}{2}$ and noting that $\sin\left(\frac{\pi}{2}\right)$ is unity while $\cos\left(\frac{\pi}{2}\right)$ vanishes, the incredible product follows:

$$\frac{2}{\pi} = \sqrt{\frac{1}{2}} \times \sqrt{\frac{1}{2}+\frac{1}{2}\sqrt{\frac{1}{2}}} \times \sqrt{\frac{1}{2}+\frac{1}{2}\sqrt{\frac{1}{2}+\frac{1}{2}\sqrt{\frac{1}{2}}}} \times \cdots.$$

§4.11. The Riemann Zeta Function

This product, called *Viète's π formula*, was discovered by the sixteenth century mathematician François Viète. Hopefully it's as phenomenal to you as we find it to be.

§4.11 The Riemann Zeta Function

From infinite products unknowingly related to π to infinite sums able to unearth the irrationality of e, our foregoing explorations into series and products have lead to many remarkable outcomes. We commence this section by recalling two of these endeavors—namely, the harmonic series (§4.3) and the Basel problem (§4.7).

Upon writing out the series concerned with each, their structural similarities become obvious:

$$\text{Harmonic series} \to \sum_{n=1}^{\infty} \frac{1}{n^1} = \frac{1}{1^1} + \frac{1}{2^1} + \frac{1}{3^1} + \frac{1}{4^1} + \cdots \tag{4.83}$$

$$\text{Basel series} \to \sum_{n=1}^{\infty} \frac{1}{n^2} = \frac{1}{1^2} + \frac{1}{2^2} + \frac{1}{3^2} + \frac{1}{4^2} + \cdots . \tag{4.84}$$

Evidently, each is of the form

$$\sum_{n=1}^{\infty} \frac{1}{n^p} \tag{4.85}$$

where p is any real number. The astute reader will recognize (4.85) as a p-series, which were briefly discussed back in §4.7. These are series *defined* by the structure of (4.85), provided $p \in \mathbb{R}$. Clearly, then, both (4.83) and (4.84) are p-series with differing p. But why should we limit p to the real numbers? Why not generalize (4.85) by extending its domain to the complex numbers $\mathbb{C}$? While at first this may appear nonsensical (what value is an arbitrary real number raised to a complex exponent?), doing so generates one of the most important functions in mathematics—the *Riemann zeta function*.

Named after the nineteenth century mathematician Bernhard Riemann, the Riemann zeta function is defined by the series

$$\zeta(s) := \sum_{n=1}^{\infty} \frac{1}{n^s} \qquad (4.86)$$

where $s \in \mathbb{C}$ and ζ is the Greek letter zeta. Obviously, whenever s is purely real we arrive back at the p-series in (4.85), but with p replaced by s. For example, setting $s = 2 + 0i = 2$ prompts the Basel series

$$\zeta(2) = \sum_{n=1}^{\infty} \frac{1}{n^2} = \frac{\pi^2}{6}. \qquad (4.87)$$

and substituting $s = 1$ yields the harmonic series

$$\zeta(1) = \sum_{n=1}^{\infty} \frac{1}{n} = \infty.$$

While this divergence at $s = 1$ may not seem interesting, it's quite an important feature to note. We will shortly arrive back at this point.

Before diving into anything really profound about (4.86), let's address the previous concern about raising arbitrary real numbers to complex exponents. By now, we are all too familiar with Euler's equation $e^{ix} = \cos(x) + i\sin(x)$. This proves that we can raise a particular real number (namely e) to a complex one. Exploiting this property, we can generalize such complex exponentiation to any positive real base z.[†] To see how, first suppose $s \in \mathbb{C}$ and so can be expressed as $s = a + bi$ where $a, b \in \mathbb{R}$ and $i := \sqrt{-1}$ is the complex unit. We write $\Re(s) = a$ to signify that the real part of s is a and $\Im(s) = b$ to reference the imaginary part b. Now suppose

---

[†] By the nature of the zeta function, we only require positive bases because the sum is over all natural numbers n.

§4.11. THE RIEMANN ZETA FUNCTION

$z^s = Q$—the value we are interested in finding. Notice

$$\begin{aligned}\log(Q) &= \log(z^{a+bi}) \\ &= \log(z^a \times z^{bi}) \\ &= \log(z^a) + \log(z^{bi}) \\ &= \log(z^a) + i\log(z^b).\end{aligned}$$

Therefore, upon exponentiating both sides we obtain

$$\begin{aligned}e^{\log(Q)} &= e^{\log(z^a) + i\log(z^b)} \\ &= e^{\log(z^a)} \times e^{i\log(z^b)}.\end{aligned}$$

Since $e^{\log(Q)} = Q$ and $e^{\log(z^a)} = z^a$, we have shown

$$Q = z^a \times e^{i\log(z^b)}.$$

But by Euler's equation, $e^{i\log(z^b)} = \cos\left(\log(z^b)\right) + i\sin\left(\log(z^b)\right)$. Consequently, for any real number $z > 0$ and any complex number $s = a + bi$,

$$z^s = z^a \left[\cos\left(\log(z^b)\right) + i\sin\left(\log(z^b)\right)\right]. \tag{4.88}$$

This shows that complex exponentiation on any positive, real base is well-defined. Hence, we need not be concerned in extending the domain of p-series to that apparent in the zeta function since the computation is quite transparent (each term in the zeta function merely assumes the form in (4.88)).

Let's now return to a larger property of the zeta function—namely, where it converges and where it diverges. As mentioned in §4.7, it is know that for any p-series with $p \leq 1$ the sum will diverge. Reasonably, then, we suspect that for all $s \in \mathbb{R}$ such that $s \leq 1$, $\zeta(s)$ will diverge too. While this is true at $s = 1$, the zeta function is suitably defined (i.e. convergent) for values $s < 1$. The reason for this is analogous to the way in which the factorial function $n!$ is well-defined for non-integer values like $n = 0$ and $n = \frac{1}{2}$. Rather than using the wonted definition of the factorial function, we extended its domain via the gamma function in §3.7. $\zeta(s)$ is similar in that its domain can be extended using a method called *analytic continuation*.[†] While too complex to describe in detail, the idea behind analytic continuation is quite simple: It extends the domain of a complex-valued function. In simpler terms, analytic continuation allows the zeta function to converge for certain values in the extended domain that would otherwise diverge in the original domain.[‡] It is important to realize that in extending via analytic continuation, the original series

$$\zeta(s) := \sum_{n=1}^{\infty} \frac{1}{n^s} \tag{4.89}$$

loses its meaning and no longer defines the zeta function. Instead, the zeta function becomes defined by the analytically continued version of (4.89), which turns out to be the Greek mess

$$\zeta(s) := 2^s \pi^{s-1} \sin\left(\frac{\pi s}{2}\right) \Gamma(1-s) \zeta(1-s), \tag{4.90}$$

where $\Gamma(1-s)$ is the gamma function with argument $1-s$. In summary, the analytically continued zeta function is the concatenation of both (4.89) and (4.90), which equates to the piecewise conglomeration

$$\zeta(s) := \begin{cases} \sum_{n=1}^{\infty} \frac{1}{n^s} & \text{if } \Re(s) > 0 \\ 2^s \pi^{s-1} \sin\left(\frac{\pi s}{2}\right) \Gamma(1-s) \zeta(1-s) & \text{otherwise.} \end{cases} \tag{4.91}$$

Indeed, claiming that the series representation for the zeta function holds for when $\Re(s) \leq 0$ is not correct. Yet it does, in some sense,

---

[†]Following analytic continuation, $s = 1$ is the only value for which $\zeta(s)$ diverges. Due to this characteristic divergence at $s = 1$, the zeta function is said to have a *singularity* or *pole* at $s = 1$.

[‡]The precise analytic continuation for the zeta function is determined by maintaining certain differentiation properties of $\zeta(s)$ in its extended domain. Under the necessary constraints, only one continuation (namely (4.90)) is possible.

§4.11. THE RIEMANN ZETA FUNCTION

allow us to assign values to these otherwise divergent series, which is what we do here.

Consider the input $s = -2k + 0i = -2k$ for some positive integer k. Clearly $\Re(s) = -2k \leq 0$. Thus, by (4.91) we have

$$\zeta(-2k) = 2^{-2k} \pi^{-2k-1} \sin(-\pi k) \Gamma(1+2k) \zeta(1+2k).$$

Though we know the sine of any integer multiple of π is zero. So it follows immediately that

$$\zeta(-2k) = 0 \qquad (4.92)$$

for all positive integers k. Arguments of this form are called the *trivial zeros* of the zeta function. "Trivial" because they are easy to come by (more on this later) and "zeros" because, well, that's obvious. Now even though the input which leads us to (4.92) is independent of the series formulation of $\zeta(s)$, (4.92) permits us to assign a value to the corresponding series. Though not logically correct, there is use in stating the following unsettling assertion

$$\sum_{n=1}^{\infty} n^{2k} = 1^{2k} + 2^{2k} + 3^{2k} + 4^{2k} + \cdots = 0 \qquad (4.93)$$

for all integers $k > 0$. In the same way taxonomy allows scientists to distinguish between different kingdoms of species, these (often bizarre) series assignments like (4.93) allow mathematicians to classify properties of zeta function-like series by their value from the analytically continued zeta function.

Another input which receives quite a bit of attention is $s = -1 + 0i = -1$. Here, $\Re(s) \leq 0$, so by (4.91)

$$\zeta(-1) = 2^{-1} \pi^{-2} \sin\left(-\frac{\pi}{2}\right) \Gamma(2) \zeta(2). \qquad (4.94)$$

As we know, $\sin\left(-\frac{\pi}{2}\right) = -1$. Moreover, from §3.7 we know for $n \in \mathbb{N}$, $\Gamma(n+1) = n!$. Hence, $\Gamma(2) = 1! = 1$. Finally, from

(4.87) we know $\zeta(2) = \frac{\pi^2}{6}$. Substituting all these results into (4.94) prompts the curious result

$$\zeta(-1) = -\frac{1}{12}.$$

Assigning this value to its corresponding series, we unearth the mysterious sum

$$\sum_{n=1}^{\infty} n = 1 + 2 + 3 + 4 + 5 + \cdots = -\frac{1}{12}. \qquad (4.95)$$

Of course, such an equality is absurd in the same way (4.93) is absurd. Nonetheless, these values allow mathematicians to assign finite values to these otherwise divergent series, which have proven to be useful in many applications, such as in the esoteric field of *string theory*[†].

A more exciting proof of (4.95) is surreptitiously obtained using a method originally conceived by the lauded mathematician Srinivasa Ramanujan. His approach goes like this: Define S_1 by the infinite series

$$S_1 := \sum_{n=0}^{\infty} (-1)^n = 1 - 1 + 1 - 1 + 1 - 1 + \cdots. \qquad (4.96)$$

In the traditional sense, this series (known as *Grandi's series* after the seventeenth-eighteenth century mathematician Luigi Guido Grandi) is a divergent one, but not in the sense that it tends off towards $\pm$ infinity. Instead, S_1 oscillates between one and zero forever, never converging to a clear value. That said, we can find a limit

---

[†]For those interested, string theory is an approach at unifying all the fundamental forces in the universe—namely: gravity, the electromagnetic force, and the two nuclear forces. String theory posits that beneath the atom and subatomic particles like quarks and gluons live tiny vibrating strings of energy that makeup all the matter in the universe. The compelling picture put forth by renowned physicist Brian Greene is a sort of "cosmic symphony" orchestrated by the music all these strings play out.

§4.11. THE RIEMANN ZETA FUNCTION

which is mildly reasonable to assign (4.96). Consider the difference $1 - S_1$, yielding

$$1 - S_1 = 1 - (1 - 1 + 1 - 1 + \cdots) = 1 - 1 + 1 - 1 + 1 - \cdots.$$

But $1 - 1 + 1 - 1 + 1 - \cdots$ is just S_1. Hence,

$$1 - S_1 = S_1 \implies S_1 = \frac{1}{2}.$$

Even though there is infinite oscillation between 1 and 0, we have shown that $S_1 = \frac{1}{2}$—the average of 0 and 1. If you think we've performed some level of mathematical wizardry, you are encouraged to visit §A.7 where we provide an additional, more rigorous method to obtain $S_1 = \frac{1}{2}$.[†]

With Grandi's series settled, we'll move on to, hopefully, a less controversial series S_2 defined by

$$S_2 := \sum_{n=1}^{\infty} n(-1)^{n-1} = 1 - 2 + 3 - 4 + 5 - 6 + 7 - 8 + \cdots. \quad (4.97)$$

This series seems straightforward enough, but does it converge? To find out, we take S_2 and add it to itself so that

$$S_2 = 0 + 1 - 2 + 3 - 4 + 5 - 6 + 7 - \cdots$$
$$+ S_2 = 1 - 2 + 3 - 4 + 5 - 6 + 7 - 8 + \cdots$$

$$2S_2 = 1 - 1 + 1 - 1 + 1 - 1 + 1 - 1 + \cdots.$$

But $S_1 = 1 - 1 + 1 - 1 + \cdots$, from which it follows that

$$2S_2 = S_1 \implies S_2 = \frac{1}{4}$$

because $S_1 = \frac{1}{2}$. We have thus shown that (4.97) converges to $\frac{1}{4}$. Now for the main punch line. We know

$$\zeta(-1) = \sum_{n=1}^{\infty} n = 1 + 2 + 3 + 4 + 5 + \cdots.$$

[†]This step is essential, so it's important you feel it is adequately justified, hence the supplement in the appendix.

Suppose we take $\zeta(-1) - 4\zeta(-1) = -3\zeta(-1)$, but add zeros in between terms so that we sum as follows:

$$\zeta(-1) = 1 + 2 + 3 + 4 + 5 + 6 + 7 + \cdots$$
$$-4\zeta(-1) = 0 - 4 + 0 - 8 + 0 - 12 + 0 - \cdots$$

$$-3\zeta(-1) = 1 - 2 + 3 - 4 + 5 - 6 + 7 - 8 + \cdots.$$

But $S_2 = 1 - 2 + 3 - 4 + 5 - 6 + \cdots$, implying

$$-3\zeta(-1) = S_2 \implies \zeta(-1) = -\frac{1}{12}$$

because $S_2 = \frac{1}{4}$. Rewriting this result, we obtain the preposterous (4.95):

$$\zeta(-1) = \sum_{n=1}^{\infty} n = 1 + 2 + 3 + 4 + 5 + 6 + 7 + \cdots = -\frac{1}{12}.$$

Again, though not a genuine equality, $-\frac{1}{12}$ is simply a useful value to assign the otherwise divergent series $1 + 2 + 3 + 4 + \cdots$.

We now take a step back for a moment only to arrive at something beyond current understanding. If you are serious about mathematics, it's possible that the zeta function is not entirely new, especially in the context of a famed problem called the *Riemann hypothesis*—one of the greatest unsolved problems in mathematics:

Conjecture 4.1 (The Riemann Hypothesis). *The nontrivial zeros of the zeta function occur with real part $\frac{1}{2}$.*

In other words, for $s \neq -2k$ where k is a positive integer (these are the trivial zeros), Riemann posits that the only other s for which $\zeta(s) = 0$ are those with $\Re(s) = \frac{1}{2}$. This is, quite literally, a million dollar question. The Clay Institute of Mathematics has attached a $1,000,000 reward for whomever cracks it first.

While this conjecture houses many important characteristics, the most significant aspect is its relationship to the obscure and

§4.11. THE RIEMANN ZETA FUNCTION

anonymous primes numbers. Though we know their are infinitely many (§4.5), primes are hard to conjure up at a moments notice. Wouldn't it be nice, for example, to have a formula that spits out all the primes less than a given value? This is one of many problems the Riemann hypothesis will help solve since the Riemann zeta function encodes information about the primes.

To unveil the link between $\zeta(s)$ and the primes $\mathbb{P}$, we turn to a discovery of Euler's. From the definition

$$\zeta(s) := 1 + \frac{1}{2^s} + \frac{1}{3^s} + \frac{1}{4^s} + \frac{1}{5^s} + \frac{1}{6^s} + \cdots$$

it follows easily that

$$\frac{1}{2^s}\zeta(s) = \frac{1}{2^s} + \frac{1}{4^s} + \frac{1}{6^s} + \frac{1}{8^s} + \frac{1}{10^s} + \frac{1}{12^s} + \cdots.$$

Taking $\zeta(s) - \frac{1}{2^s}\zeta(s)$ we have

$$\zeta(s) - \frac{1}{2^s}\zeta(s) = \left(1 + \frac{1}{2^s} + \frac{1}{3^s} + \frac{1}{4^s} + \cdots\right)$$
$$- \left(\frac{1}{2^s} + \frac{1}{4^s} + \frac{1}{6^s} + \frac{1}{8^s} + \cdots\right).$$

This rearranges to

$$\zeta(s)\left(1 - \frac{1}{2^s}\right) = 1 + \frac{1}{3^s} + \frac{1}{5^s} + \frac{1}{7^s} + \frac{1}{9^s} + \cdots. \quad (4.98)$$

Notice that on the right-hand side of (4.98), all the integers in the denominators are odd. Suppose now we are interested in the series given by $\frac{1}{3^s}\left(1 - \frac{1}{2^s}\right)\zeta(s)$. From (4.98) we can write

$$\frac{1}{3^s}\left(1 - \frac{1}{2^s}\right)\zeta(s) = \frac{1}{3^s} + \frac{1}{9^s} + \frac{1}{15^s} + \frac{1}{21^s} + \cdots,$$

which implies

$$\zeta(s)\left(1 - \frac{1}{2^s}\right) - \frac{1}{3^s}\zeta(s)\left(1 - \frac{1}{2^s}\right)$$
$$= \left(1 + \frac{1}{3^s} + \frac{1}{5^s} + \cdots\right) - \left(\frac{1}{3^s} + \frac{1}{9^s} + \frac{1}{15^s} + \cdots\right).$$

A simple algebraic reformulation prompts

$$\zeta(s)\left(1-\frac{1}{2^s}\right) - \frac{1}{3^s}\zeta(s)\left(1-\frac{1}{2^s}\right) = \zeta(s)\left(1-\frac{1}{2^s}\right)\left(1-\frac{1}{3^s}\right)$$

yielding

$$\zeta(s)\left(1-\frac{1}{2^s}\right)\left(1-\frac{1}{3^s}\right) = 1+\frac{1}{5^s}+\frac{1}{7^s}+\frac{1}{11^s}+\frac{1}{13^s}+\cdots. \quad (4.99)$$

Similar to before, notice how all the integers in the denominators of (4.99) are neither powers of two nor three. In essence, we have removed all the prime factors of two and three from the right-hand side of the zeta function. In general, repeating this process up to the kth prime p_k will remove all the numbers with a prime factor of p_k on the right-hand side. Doing so forever removes all the fractions on the right-hand side, leaving only the paramount integer one.

Symbolically, this infinite procedure would manufacture the expression

$$\zeta(s)\left(1-\frac{1}{2^s}\right)\left(1-\frac{1}{3^s}\right)\left(1-\frac{1}{5^s}\right)\left(1-\frac{1}{7^s}\right)\cdots = 1,$$

which can be reformulated so that

$$\zeta(s) = \left(\frac{2^s}{2^s-1}\right)\left(\frac{3^s}{3^s-1}\right)\left(\frac{5^s}{5^s-1}\right)\left(\frac{7^s}{7^s-1}\right)\cdots.$$

More succinctly, we have

$$\zeta(s) = \prod_{k=1}^{\infty}\left(\frac{p_k^s}{p_k^s-1}\right), \quad (4.100)$$

where p_k is the kth prime number. It should now be obvious that the primes and the zeta function are intimately related, and it is this relationship that makes the Riemann zeta function such an important one. That said, there is more to (4.100) and the connection between $\zeta(s)$ with $\mathbb{P}$ than meets the eye.

There is a very famous theorem in mathematics called the *prime number theorem* (PNT) that asserts the number of primes less than

§4.11. The Riemann Zeta Function

a positive number n is approximately $\frac{n}{\log(n)}$. Less naïvely, the PNT states

$$\pi(n) \sim \frac{n}{\log(n)}. \tag{4.101}$$

Here, $\pi(n)$ is coined the *prime counting function*—a function that computes the number of primes less than the input $n > 1$. The notation $f(n) \sim g(n)$ denotes an *asymptotic relationship* between the two function $f(n)$ and $g(n)$. It means that as $n \to \infty$, $f(n)$ and $g(n)$ become equal. In other words,

$$f(n) \sim g(n) \implies \lim_{n \to \infty} \frac{f(n)}{g(n)} = 1.$$

In the context of (4.101), the notation signifies

$$\lim_{n \to \infty} \frac{\pi(n)}{\frac{n}{\log(n)}} = \lim_{n \to \infty} \frac{\pi(n) \log(n)}{n} = 1.$$

This is to say that the error in $\pi(n)$ tends toward zero as n increases, which implies that as $n \to \infty$, $\pi(n)$ approaches the exact number of primes less than n, and converges to the ratio $\frac{n}{\log(n)}$ for sufficiently large n.

What's incredible about $\pi(n)$ is its margin of error in counting the number of primes is closely tied to the distribution of the zeta function's nontrivial zeros. So an understanding of the zeta function lends insight into the prime-counting function, and thus into the distribution of the prime numbers. Diagrammatically,

$$\zeta(s) \longrightarrow \pi(n) \longrightarrow \mathbb{P},$$

where each $\longrightarrow$ should be guilelessly read "tells me about." This is something all number theorists, if not all mathematicians, fervently desire to understand.

Chapter 5

Beyond Calculus

"God made the integers, all the rest is the work of man."
 ~ Leopold Kronecker

ALCULUS is an omnipotent mathematical tool. Some find pleasure in its formulation, and nearly all find it useful. That said, calculus is but a planet in the whole galaxy of mathematics. There are many other fields within mathematics just as invigorating as calculus, and it is the purpose of this chapter to describe a few of them in the context of a problem characteristic to each field.

§5.1   Algebra: Nested Radicals and Tetration

Algebra is the basis of calculus and much of mathematics. Without it, nearly all the ideas established throughout this book would be unfounded. Yet even with high importance, many associate algebra with aimless arithmetic and tedious equation reformulation. While these are substantiated at times, they should not submit to defining algebra itself. Instead, algebra should remain defined as a field in foundational mathematics whose versatility is as astounding as some of the problems in the field. Here we would like to discuss two

§5.1. ALGEBRA: NESTED RADICALS AND TETRATION

particular problems whose rudimentary solutions are suspiciously so, given the apparent complexity of the problem.

The first concerns an algebraic structure called an *iterated* or *nested radical* (we prefer the latter).[†] These, as the names suggest, are repetitive square roots. For instance, the value $\sqrt{\sqrt{3}}$ could be described as a doubly-iterated or doubly-nested radical because there are are two consecutive square roots being applied. Of interest to us are not finite nested radicals, but infinite nested radicals—namely, the expression:

$$\sqrt{1 + 2\sqrt{1 + 3\sqrt{1 + 4\sqrt{1 + \cdots}}}}. \tag{5.1}$$

Our question is simple: To what value does this converge?

Originally posed in 1911 by the prominent mathematician Srinivasa Ramanujan, this question has a notorious history of being deceptively difficult. That is, unless an adequate formulation of the pattern is unveiled, from which the elementary solution follows immediately.[‡]

We will begin our evaluation of (5.1) by considering the quadratic expression

$$f(n) = n(n+2).$$

Such a function may seem arbitrary, but it is quite fundamental to the problem at hand. Observe the rather involved algebraic rear-

---

[†] Recall that we encountered a version of nested radicals in §4.10.
[‡] The reader is encouraged to attempt the evaluation of (5.1) before reading our analysis.

rangement:

$$\begin{aligned}f(n) &= n(n+2)\\&=n\sqrt{(n+2)^2}\\&=n\sqrt{n^2+4n+4}\\&=n\sqrt{1+(n^2+4n+3)}\\&=n\sqrt{1+(n+1)(n+3)}.\end{aligned}$$

Note further that the product

$$(n+1)(n+3) = f(n+1).$$

Hence, we have shown

$$f(n) = n\sqrt{1+f(n+1)}.$$

Such a formulation may seem useless, but upon plugging in $n=1$ a magical thing happens:

$$\begin{aligned}f(1) &= \sqrt{1+f(2)}\\&=\sqrt{1+2\sqrt{1+f(3)}}\\&=\sqrt{1+2\sqrt{1+3\sqrt{1+f(4)}}}\\&=\sqrt{1+2\sqrt{1+3\sqrt{1+4\sqrt{1+\cdots}}}}.\end{aligned}$$

This is the original nested radical (5.1)! Thus we have shown $f(1)$ is equal to (5.1). And because $f(n) = n(n+2)$, we have that $f(1) = 3$, from which is follows that

$$3 = \sqrt{1+2\sqrt{1+3\sqrt{1+4\sqrt{1+\cdots}}}}.$$

As an aside, we note that once given the solution to (5.1), its

§5.1. Algebra: Nested Radicals and Tetration

verification follows immediately:

$$\begin{aligned}
3 &= \sqrt{9} \\
&= \sqrt{1 + 2(4)} \\
&= \sqrt{1 + 2\sqrt{16}} \\
&= \sqrt{1 + 2\sqrt{1 + 3(5)}} \\
&= \sqrt{1 + 2\sqrt{1 + 3\sqrt{25}}} \\
&= \sqrt{1 + 2\sqrt{1 + 3\sqrt{1 + 4(6)}}} \\
&= \sqrt{1 + 2\sqrt{1 + 3\sqrt{1 + 4\sqrt{36}}}} \\
&= \sqrt{1 + 2\sqrt{1 + 3\sqrt{1 + 4\sqrt{1 + \cdots}}}}.
\end{aligned}$$

Our next problems concerns another familiar algebraic operator ad infinitum—that of exponentiation. We are accustomed to algebraic equations like 2^x and certainly the proverbial e^x. Here, however, we are interested in neither of these singly-exponentiated functions, and instead want to consider expressions similar to

$$x^x, x^{x^x}, x^{x^{x^x}}, \ldots .$$

The process of forming these iterated exponential functions is called *tetration*.[†] Tetration is the operation that follows exponentiation, similar to how exponentiation follows multiplication, and how multiplication follows addition.

To illustrate the tetration process, recall that a number a exponentiated to n, denoted a^n, is given by

$$a^n = \underbrace{a \times a \times \cdots \times a}_{n \text{ times}}.$$

[†] Tetration is sometimes referred to as *hyper-4* and a tetrated expression a *power tower*.

Correspondingly, for a tetrated to n, denoted ${}^n a$, we have

$$ {}^n a = \underbrace{a^{a^{\cdot^{\cdot^{\cdot^a}}}}}_{n \text{ times}}. $$

And similar to how you can take infinite exponentials

$$ \lim_{n \to \infty} x^n = x \times x \times x \times \cdots, $$

we are interested in infinite tetrationals[†] such as

$$ \lim_{n \to \infty} {}^n x = x^{x^{x^{\cdot^{\cdot^{\cdot}}}}}. $$

In particular, our question is concerned with the infinite tetration

$$ 2 = {}^\infty x = x^{x^{x^{\cdot^{\cdot^{\cdot}}}}}. \tag{5.2} $$

We inquire: For what x will (5.2) be true?

Similar to the nested radicals above (which are really just a special case of tetration), we are interested in deducing some recursive structure in the problem. There are a multitude of ways to see this, but perhaps the most obvious is to take the logarithm with base x, denoted $\log_x$, of both sides:

$$ \log_x(2) = \log_x\left(x^{x^{x^{\cdot^{\cdot^{\cdot}}}}} \right). $$

By the logarithm property $\log(a^b) = b \log(a)$, this is equivalent to the equality

$$ \log_x(2) = \left(x^{x^{\cdot^{\cdot^{\cdot}}}} \right) \log_x(x) = x^{x^{x^{\cdot^{\cdot^{\cdot}}}}} = x^{x^{x^{\cdot^{\cdot^{\cdot}}}}} $$

where in the second step we used the fact that $\log_x(x) = 1$ for all x. Notice the quantity

$$ x^{x^{x^{\cdot^{\cdot^{\cdot}}}}} = 2 $$

[†]The word "tetrational" is probably a hypercorrection, but we cannot say for sure.

from (5.2) above. Therefore, we have reduced the problem down to solving
$$\log_x(2) = 2 \implies 2 = x^2,$$
whose solution is clearly $x = \pm\sqrt{2}$. Because we have taken the logarithm of x, we require $x > 0$. Hence, we have proved the solution to the equation
$$2 = x^{x^{x^{\cdot^{\cdot^{\cdot}}}}}$$
is $x = \sqrt{2}$. As intractable a problem this may have seemed, its calculation turned out to be quite straightforward. Indeed, even the monotonous field of algebra incorporates a few surprises.

§5.2 Number Theory: The Golden Ratio

First encountered in §4.5, the field of *number theory* is concerned with the properties and structure of the real numbers. To quote Gauss, "Mathematics is the queen of the sciences and number theory is the queen of mathematics." Our exploration into number theory will be in studying a famous sequence of numbers and its intrinsic relationship to the number $1.618033\cdots$.

As you may know, this number is the *golden ratio*. Symbolically, it is denoted φ and specified by
$$\varphi = \frac{1+\sqrt{5}}{2} = 1.618033\cdots.$$

Of course, just stating the value of φ does it little justice in explaining its innate link to the integers. To procure some interest into φ, let's construct a sequence of numbers with the following rules:

- The first and second numbers in the sequence are one.
- The next number is equal to the sum of the previous two in the sequence.

If F_n is the nth element in the sequence, then these rules are notated as thus:

- $F_1 = F_2 = 1$.
- $F_n = F_{n-1} + F_{n-2}$.

Using these rules, we can generate the first few numbers: $F_3 = 1 + 1 = 2$, $F_4 = 1 + 2 = 3$, $F_5 = 2 + 3 = 5$, and so forth. Doing so for many iterations produces the sequence

$$1, 1, 2, 3, 5, 8, 13, 21, 34, 55, 89, 144, 233, 377, 610 \cdots. \quad (5.3)$$

This sequence of numbers is called the *Fibonacci sequence*, named after the thirteenth century mathematician Leonardo Pisano Bigollo (a.k.a. Fibonacci). Of interest here are not the elements in (5.3) themselves, but rather the ratio between consecutive elements. Consider, for instance, the ratio between the fifth and fourth Fibonacci numbers:

$$\frac{F_5}{F_4} = \frac{5}{3} = 1.66\overline{6}.$$

There is nothing too exciting here, but performing such a computation on larger Fibonacci numbers yields a rather suspicious result:

$$\frac{F_{15}}{F_{14}} = \frac{610}{377} = 1.618037\cdots \approx \varphi.$$

Such an observation prompts the following inquiry: Could it be that

$$\lim_{n \to \infty} \frac{F_{n+1}}{F_n} = \varphi? \quad (5.4)$$

To address this question, recall the primary rule of the Fibonacci sequence—that $F_n = F_{n-1} + F_{n-2}$. With n increased by one, this rule is just $F_{n+1} = F_n + F_{n-1}$. For sufficiently large n (that is, large enough n such that we need not consider the limit $n \to \infty$), consider the ratio in (5.4). We define

$$\phi := \frac{F_{n+1}}{F_n} = \frac{F_n}{F_{n-1}}.$$

§5.2. Number Theory: The Golden Ratio

By the rule above, it follows that

$$\phi = \frac{F_n + F_{n-1}}{F_n} = \frac{F_n}{F_{n-1}} \implies 1 + \frac{F_{n-1}}{F_n} = \frac{F_n}{F_{n-1}}.$$

But the ratio $\frac{F_{n-1}}{F_n} = \frac{1}{\phi}$ because, by definition, $\phi := \frac{F_n}{F_{n-1}}$. Substituting this in, we obtain the quadratic expression

$$1 + \frac{1}{\phi} = \phi \implies \phi^2 - \phi - 1 = 0. \tag{5.5}$$

Solving for ϕ with the quadratic formula, we find

$$\phi = \frac{1 \pm \sqrt{5}}{2}. \tag{5.6}$$

And because $F_n > 0$ for all n, we expect the ratio $\frac{F_{n+1}}{F_n} > 0$. Therefore, we take the positive solution in (5.6) and conclude that

$$\phi = \frac{1 + \sqrt{5}}{2} = \varphi,$$

confirming our theory that the ratio between consecutive Fibonacci numbers converges to the golden ratio.

But there is more to this proof than meets the eye. Nowhere throughout the derivation did we use the first defining feature of the Fibonacci sequence—that $F_1 = 1$ and $F_2 = 1$. Thus, not only does the ratio of consecutive Fibonacci numbers converge to the golden ratio, but the ratio of consecutive integers in *all sequences* defined by the rule that the next element is the sum of the previous two converge to the golden ratio. In other words, so long $F_n = F_{n-1} + F_{n-2}$, $\frac{F_{n+1}}{F_n} \to \varphi$ as $n \to \infty$ irrespective of F_1 and F_2.

To promote more interest into the golden ratio, we provide another means to compute it using nested radicals. Consider

$$z = \sqrt{1 + \sqrt{1 + \sqrt{1 + \sqrt{1 + \cdots}}}}.$$

Taking z^2 we obtain

$$z^2 = 1 + \sqrt{1 + \sqrt{1 + \sqrt{1 + \cdots}}}.$$

But the quantity
$$\sqrt{1+\sqrt{1+\sqrt{1+\cdots}}} = z,$$
implying
$$z^2 = 1+z \implies z^2 - z - 1 = 0.$$
This is the same polynomial as in (5.5), meaning
$$z = \frac{1 \pm \sqrt{5}}{2}.$$
And because the square root yields a positive value, we must have that
$$z = \frac{1+\sqrt{5}}{2} = \varphi.$$
Thus we have shown the golden ratio
$$\varphi = \sqrt{1+\sqrt{1+\sqrt{1+\sqrt{1+\cdots}}}}.$$

As exciting as this iterated radical is, we would like to end this section again with the Fibonacci sequence. Rather than list off the sequence through the recurrence relationship $F_n = F_{n-1} + F_{n-2}$, we would like a formula that generates F_n without knowledge of F_{n-1} or F_{n-2}. But instead of solving the recurrence relationship $F_n = F_{n-1} + F_{n-2}$ as we've done previously, we intend to prove that the formula
$$F_n = \frac{1}{\sqrt{5}} \left(\varphi^n - \xi^n \right),$$
where $\xi = \frac{1-\sqrt{5}}{2}$ is the conjugate of the golden ratio, is a formula for the nth Fibonacci number. This formula, by the way, is called *Binet's formula* after the nineteenth century mathematician Jacques Binet. We'll prove his formula using a proof technique called *induction*, which we now discuss.

A proof by induction allows us to verify a proposed formula whose domain is, for our purposes, the set of natural (or whole)

§5.2. Number Theory: The Golden Ratio

numbers. An inductive proof always begins by proving the *base case*, which is simply showing that the proposed formula works for the smallest value in the domain (e.g. $n = 1$). We then make the *inductive hypothesis*—the assumption that the proposed formula holds for an arbitrary number n in the domain. Then, using this assumption and the base case, we prove that the formula holds for the next number in the domain—namely $n + 1$. Why this proves the formula should be convincing: We showed the formula works for $n = 1$ and also $n + 1$. Hence, it must work for $n = 2$. But it works for $n + 1$, so it must work for $n = 3$. But it works for $n + 1$, so it must work for $n = 4$. This continues, extending to all values in the domain.[†]

We'll now illustrate a proof by induction in substantiating the following lemma:

Lemma 5.1. *If x is such that $x^2 = x + 1$, then*

$$x^{n+1} = xF_{n+1} + F_n, \qquad (5.7)$$

where F_n is the nth Fibonacci number.

Proof. We start by acknowledging the domain of F_n. We saw previously that the smallest Fibonacci number is $F_1 = 1$. Hence, the domain to which (5.7) applies is for integers $n \geq 1$. That is, $n \in \mathbb{N}$. Consequently, the base case of the inductive proof is for $n = 1$, which is showing $x^2 = x + 1$ because $F_2 = F_1 = 1$. Inputting $n = 1$, we obtain

$$x^2 = xF_2 + F_1$$
$$= x(1) + 1$$
$$= x + 1.$$

Therefore, the formula holds for $n = 1$ and so the base case is proven.

[†]A more comprehensive description and additional examples of induction are provided in §A.8.

The next step is to make the inductive hypothesis. Here, we assume (5.7) is true for some n. Under this assumption (and perhaps use of the base case), proving (5.7) holds for $n+1$ proves it works for all $n \in \mathbb{N}$. Notice that for $n+1$, $x^{(n+1)+1} = x^{n+2} = x \times x^{n+1}$. But from our inductive hypothesis we know $x^{n+1} = xF_{n+1} + F_n$. Hence

$$x^{n+2} = x\left(xF_{n+1} + F_n\right)$$
$$= x^2 F_{n+1} + xF_n.$$

In the statement of Lemma 5.1, we required x to be such that $x^2 = x+1$. Therefore,

$$x^{n+2} = x^2 F_{n+1} + xF_n$$
$$= (x+1)F_{n+1} + xF_n$$
$$= x(F_{n+1} + F_n) + F_{n+1}.$$

But by the definition of the Fibonacci sequence, $F_{n+2} = F_{n+1} + F_n$. Thus, $x^{n+2} = xF_{n+2} + F_{n+1}$, which matches (5.7) with input $n+1$, and so the proof is complete. □

Perhaps it's not exactly clear why this lemma will be of use. Recall our restriction on what x had to satisfy in order for (5.7) to be used: x must be such that $x^2 = x+1$. But as we saw previously, the only x which satisfy this quadratic are

$$x = \frac{1 \pm \sqrt{5}}{2};$$

that is, both φ and ξ. Hence, from Lemma 5.1 with input $n-1$, we have shown the following two results:

$$\varphi^n = \varphi F_n + F_{n-1} \tag{5.8}$$

$$\xi^n = \xi F_n + F_{n-1}. \tag{5.9}$$

Solving for F_{n-1} in (5.8), we have

$$F_{n-1} = \varphi^n - \varphi F_n.$$

Substituting this into (5.9), we obtain

$$\xi^n = \xi F_n + \varphi^n - \varphi F_n \implies F_n = \frac{\varphi^n - \xi^n}{\varphi - \xi}.$$

But

$$\varphi - \xi = \frac{1+\sqrt{5}}{2} - \frac{1-\sqrt{5}}{2} = \sqrt{5}.$$

Consequently, we have proven Binet's remarkable formula

$$F_n = \frac{1}{\sqrt{5}} \left(\varphi^n - \xi^n \right) \qquad (5.10)$$

for the nth Fibonacci number. It is quite satisfying to see this formula generate the Fibonacci numbers, particularly to see the $\frac{1}{\sqrt{5}}$ cancel in each case. You are encouraged to generate the first few using (5.10).

§5.3   Combinatorics: Pascal's Triangle

Combinatorics is the study of counting and the many ingenious ways of doing so. Consider, for instance, our friend the factorial function. In §3.7, we studied (for $n \in \mathbb{N}$) the relationship between

$$n! := n \times (n-1) \times \cdots \times 2 \times 1$$

and the number of way to arrange n objects, such as the letters ABC:

$$\underbrace{ABC, ACB, BAC, BCA, CAB, CBA}_{3! \,=\, 6 \text{ ways.}}.$$

This is one of many functions that vastly shortens the task of determining the number of possible arrangements, combinations, or permutations a set of objects can presume. In this section, we go a bit beyond the factorial function and introduce the binomial coefficient and its relationship to mathematician Blaise Pascal's famous triangle.

By definition, the *binomial coefficient* for two integers n and m such that $n \geq m \geq 0$ is

$$_nC_k = \binom{n}{k} = \frac{n!}{k!(n-k)!}. \qquad (5.11)$$

The notations $_nC_k$ and $\binom{n}{k}$ are read "n choose k"[†] because the binomial coefficient measures the number of ways to choose k items from a set of n objects (without replacement and irrespective of order).

To illustrate, suppose we want to know how many ways there are to choose two cards from a deck of fifty-two (without replacement and without consideration for the order in which the cards are drawn). Our first choice involves choosing one of the starting fifty-two cards, from which there are fifty-two ways to do so. Suppose we choose the ace of clubs. Now the deck has fifty-one cards, from which there are fifty-one ways to choose the next card. Suppose our next choice is the two of hearts. You may be inclined to multiply these two values together and conclude that there are $52 \times 51 = 2652$ ways to choose the two cards. This answer, however, fails to account for the requirement that the drawing order be immaterial. The proper number is obtained by dividing out the number of ways two cards can be arranged (why?). Recalling our discussion from §3.7, the number of ways n objects can be arranged is $n!$. Hence, the number of ways two cards can be arranged is $2! = 2$. Our final answer to the above puzzle is thus

$$\frac{52 \times 51}{2!} = \frac{2656}{2!} = 1326 \text{ ways.}$$

[†]We hereafter use the latter notation.

§5.3. COMBINATORICS: PASCAL'S TRIANGLE

Unsurprisingly, using (5.11) above produces an identical result:

$$\binom{52}{2} = \frac{52!}{2!(52-2)!}$$
$$= \frac{52 \times 51 \times 50 \times \cdots \times 2 \times 1}{2!(50 \times \cdots \times 2 \times 1)}$$
$$= \frac{52 \times 51}{2!}$$
$$= 1326 \text{ ways.}$$

The previous analysis provides a means for justifying why (5.11) is the correct formula for the number of ways to choose k objects from a set of n. Instead of a fifty-two card deck, consider the more general deck with n cards. We wish to compute the number of ways to choose k of these cards (importantly, without replacement and irrespective of the order in which the cards are drawn). There are n ways to choose the first card, $n-1$ to choose the second, $n-2$ to choose the third and so forth, up to the $(n-(k+1))$th ways to choose the kth card. In total, there are

$$\frac{n!}{(n-k)!} = n(n-1)(n-2)\cdots(n-(k+1)) \qquad (5.12)$$

ways to choose k cards from a deck of n without replacement. As with the two cards above, to remove all instances of order bias we simply divide (5.12) by the number of ways the k cards can be arranged, $k!$. Hence,

$$\binom{n}{k} = \frac{n!}{k!(n-k)!}.$$

To apply this new-founded idea of choosing k objects from a set of n, we pursue a popular combinatorics problem concerning the sum

$$\sum_{k=0}^{n} \binom{n}{k}$$

and the rows of *Pascal's triangle*—a triangular table whose first five rows are:

$n=0$					1			
$n=1$				1		1		
$n=2$			1		2		1	
$n=3$		1		3		3		1
$n=4$	1		4		6		4		1

The construction of the next row is quite simple. The end points are always one, while those in between are the sum of the two terms directly above. This makes the next row ($n=5$)

$$1 \quad 5 \quad 10 \quad 10 \quad 5 \quad 1$$

This triangle has a plethora of fascinating properties. One of the more interesting being that the sum of the values on the nth row is 2^n. More mathematically, if $\mathcal{P}_n$ is the set of elements on the nth row of Pascal's triangle, then

$$\sum_{j \in \mathcal{P}_n} j = 2^n, \tag{5.13}$$

where the sum is taken over all elements j in the set $\mathcal{P}_n$. This we use to motivate an upcoming problem.

Not at all obvious is the relationship between Pascal's triangle and binomial coefficients. Only after computing many binomial coefficients will you have the chance of stumbling across the pattern that a triangle of binomial coefficients (hereafter called the *binomial triangle*) equates to Pascal's triangle:

§5.3. COMBINATORICS: PASCAL'S TRIANGLE

$n = 0$ $\quad\binom{0}{0}$

$n = 1$ $\quad\binom{1}{0} \quad \binom{1}{1}$

$n = 2$ $\quad\binom{2}{0} \quad \binom{2}{1} \quad \binom{2}{2}$

$n = 3$ $\quad\binom{3}{0} \quad \binom{3}{1} \quad \binom{3}{2} \quad \binom{3}{3}$

$n = 4$ $\quad\binom{4}{0} \quad \binom{4}{1} \quad \binom{4}{2} \quad \binom{4}{3} \quad \binom{4}{4}$

Of course, if the binomial triangle is identical to Pascal's, then the additive relationships among the rows of Pascal's triangle must hold for the binomial version as well. For example, at the fourth and fifth rows ($n = 3, 4$) in the binomial triangle, you can verify that indeed

$$\binom{4}{2} = \binom{3}{1} + \binom{3}{2}.$$

But in order for the binomial triangle to equate exactly to Pascal's, we require this to be true for all rows. This leads us to the following relation

$$\binom{n+1}{k} = \binom{n}{k-1} + \binom{n}{k}.$$

This formula is called *Pascal's identity* and is proved in the following lemma:

Lemma 5.2. *If $n, k \in \mathbb{N}_0$ such that $n \geq k$, then*

$$\binom{n+1}{k} = \binom{n}{k-1} + \binom{n}{k}. \tag{5.14}$$

Proof. We will prove (5.14) using the factorial definition of the binomial coefficient described previously in (5.11). Using this definition, showing (5.14) is equivalent to showing

$$\frac{(n+1)!}{k!(n+1-k)!} = \frac{n!}{k!(n-k)!} + \frac{n!}{(k-1)!(n-k+1)!}. \tag{5.15}$$

A little algebra brings us to the required result:

$$\frac{n!}{k!(n-k)!} + \frac{n!}{(k-1)!(n-k+1)!} = \frac{n!(n-k+1) + n!(k)}{k!(n-k+1)!}$$
$$= \frac{n![(n+1)-k] + n!(k)}{k!(n+1-k)!}$$
$$= \frac{(n+1)!}{k!(n+1-k)!}.$$

This verifies (5.15), and so the proof is complete. □

The proof of this identity not only verifies that the binomial triangle is identical to Pascal's, but also leads to the sum we referenced earlier. We observed with Pascal's triangle (5.13)—that the sum of the values on the nth row of Pascal's triangle is 2^n. Since Pascal's triangle is identical to the binomial triangle, the same result should hold for the binomial triangle (assuming (5.13) is genuine). In terms of the binomial triangle, the conjectured sum is

$$\sum_{k=0}^{n} \binom{n}{k} = 2^n. \tag{5.16}$$

It is the purpose of this section to prove (5.16) using two separate techniques—the first uses algebra, and the second combinatorics.

To prove with algebra, we use an inductive argument with base case $n = 0$ (the minimum value in the domain of (5.16)). We have

$$\sum_{k=0}^{0} \binom{n}{k} = \binom{0}{0} = 1 = 2^0,$$

which is the crest of both the binomial and Pascal's triangle. Our inductive hypothesis is to assume (5.16) holds for some value n. We will prove it works for $n + 1$.

We have

$$\sum_{k=0}^{n+1} \binom{n+1}{k} = \underbrace{\binom{n+1}{0} + \binom{n+1}{n+1}}_{\text{these are the first and last terms}} + \sum_{k=1}^{n} \binom{n+1}{k}.$$

§5.3. Combinatorics: Pascal's Triangle

Here, $\binom{n+1}{0} = \binom{n+1}{n+1} = 1$. Applying Pascal's identity to the summand, we obtain

$$\binom{n+1}{0} + \binom{n+1}{n+1} + \sum_{k=1}^{n}\binom{n+1}{k}$$
$$= 2 + \sum_{k=1}^{n}\left[\binom{n}{k-1} + \binom{n}{k}\right]$$
$$= 2 + \sum_{k=1}^{n}\binom{n}{k-1} + \sum_{k=1}^{n}\binom{n}{k}.$$

But from the inductive hypothesis,

$$2^k = \sum_{k=0}^{n}\binom{n}{k} = \binom{n}{n} + \sum_{k=1}^{n}\binom{n}{k-1}$$
$$\implies \sum_{k=1}^{n}\binom{n}{k-1} = 2^k - 1.$$

Likewise,

$$2^k = \sum_{k=0}^{n}\binom{n}{k} = \binom{n}{0} + \sum_{k=1}^{n}\binom{n}{k}$$
$$\implies \sum_{k=1}^{n}\binom{n}{k} = 2^k - 1.$$

Therefore, we have

$$2 + \sum_{k=1}^{n}\binom{n}{k-1} + \sum_{k=1}^{n}\binom{n}{k} = 2 + (2^k - 1) + (2^k - 1)$$
$$= 2^k(1+1)$$
$$= 2^{k+1},$$

which is to say

$$\sum_{k=0}^{n+1}\binom{n+1}{k} = 2^{k+1}.$$

This is (5.16) with upper limit $n+1$, which is the sum we are looking for. Hence, (5.16) is proved.

This was quite an algebraic trudge, and may be steering you away from enjoying combinatorics. But we have only just arrived at the combinatorial approach in proving (5.16), so do not be discouraged. This proof is must shorter, and requires us to think about what $\binom{n}{k}$ actually means (combinatorially speaking).

Recall that $\binom{n}{k}$ computes the number of ways to take k objects (without replacement) from a set of n without regard for the order in which the objects are removed. The number of ways to choose *any* number of objects from the set of n is the number of ways to choose zero objects plus the number of ways to choose one object plus the number of ways to choose two objects and so forth. In other words, the number of ways to make a subset of objects from a set of n is the familiar sum

$$\binom{n}{0} + \binom{n}{1} + \binom{n}{2} + \cdots + \binom{n}{n} = \sum_{k=0}^{n} \binom{n}{k}. \qquad (5.17)$$

But let's approach this from a separate perspective. If we have a set of n objects, the number of ways to choose any number of these objects is contingent upon the binary nature of choosing an object— either we choose it or not. Hence, for each object there are two possible outcomes—it is either in our subset or not. There are n total objects, so the total number of possible choices is

$$\underbrace{2 \times 2 \times 2 \times \cdots \times 2}_{n \text{ times}}.$$

This is 2^n—the number of ways to make a subset of objects from a set of n. But this is precisely the problem to which (5.17) provides an answer. The only logical conclusion is that these two results are equal. Consequently, we have shown through a combinatorial argument that

$$\sum_{k=0}^{n} \binom{n}{k} = 2^n.$$

The logics of this proof are certainly more entertaining than the tedious algebraic one, which is why many who enjoy mathematical reasoning end up studying combinatorics. Of course, if sterling logical arguments are what you fancy, then set theory—the topic of the next section—is the field for you.

§5.4 Set Theory: $\infty > \infty$

Set theory is the logical foundation on which much of mathematics rests. It is an exploration into the foundational logics of mathematics and also a rigorous analysis of sets, set operations, and their applications to more well-known fields of mathematics such as calculus.

Our venture into set theory will be in exploring a more rigorous interpretation of infinity and infinite sets, such as $\mathbb{R}$ and $\mathbb{Z}$. We aspire to prove that some infinities are literally larger than others. To do this, we require some additional language and notation omnipresent throughout the field.

First, it is best to recall the definition of a set, originally encountered at the beginning of the book: A *set* is nothing but a collection of distinct objects. Whether composed of numbers, letters, or Russian symbols, the elements of the set do not change the fact that the overarching mathematical construction is a set.

A natural property of a set is its size, that is how many elements are contained within the set. Mathematicians call the size of a set its *cardinality*. Like the absolute value of a number, the cardinality of a set is represented using vertical bars $|\ |$. For example, the set

$$\mathcal{B} = \{1, 11, 21, 1211, 111221\}$$

has cardinality five because there are five elements in $\mathcal{B}$. In notation, we write $|\mathcal{B}| = 5$. Similarly, as we mentioned above, the sets $\mathbb{R}$ and $\mathbb{Z}$ are infinite sets, meaning $|\mathbb{R}| = \infty$ and $|\mathbb{Z}| = \infty$. The *null set*, denoted $\varnothing$, is a set whose cardinality, by definition, is zero. Hence, $|\varnothing| = 0$.

Like any mathematical object, it would be useful to introduce some notation that allows us to manipulate and compare sets. In other words, we would like some sort of "set-arithmetic," so to speak. Of all possible *set operations*, the two most common are the *union* and *intersection* of two sets. The union of the sets $\mathcal{M}$ and $\mathcal{N}$, denoted $\mathcal{M} \cup \mathcal{N}$,[†] is the set of elements *in either* $\mathcal{M}$ and $\mathcal{N}$. Conversely, the intersection of $\mathcal{M}$ and $\mathcal{N}$, denoted $\mathcal{M} \cap \mathcal{N}$, is the set of elements *in both* $\mathcal{M}$ and $\mathcal{N}$.

To illustrate, suppose

$$\mathcal{M} = \{6, 28, 496\} \text{ and } \mathcal{N} = \{1, 3, 6\}.$$

It follows that the union of $\mathcal{M}$ and $\mathcal{N}$ is $\{1, 3, 6, 28, 496\}$, that is

$$\mathcal{M} \cup \mathcal{N} = \{1, 3, 6, 28, 496\},$$

because these are all the unique elements between the sets $\mathcal{M}$ and $\mathcal{N}$. On the other hand, the intersection of $\mathcal{M}$ and $\mathcal{N}$ is $\{6\}$, that is

$$\mathcal{M} \cap \mathcal{N} = \{6\},$$

since six is the only element in common to both sets.

As an aside, notice that for the null set $\varnothing$ and for any other set $\mathcal{A}$ with all distinct elements,

$$\mathcal{A} \cup \varnothing = \mathcal{A}$$

while

$$\mathcal{A} \cap \varnothing = \varnothing.$$

Like series and products, we may want to express many unions at once. To do this, we employ the sum- and product-like notation

$$\bigcup_{n=1}^{p} \mathcal{V}_n = \mathcal{V}_1 \cup \mathcal{V}_2 \cup \cdots \cup \mathcal{V}_p,$$

[†]Remember $\cup$ for *Union*.

§5.4. SET THEORY: $\infty > \infty$

where $\mathcal{V}_1, \mathcal{V}_2, \cdots, \mathcal{V}_p$ are arbitrary sets. Analogous to unions, for many intersection we would write

$$\bigcap_{n=1}^{p} \mathcal{V}_n = \mathcal{V}_1 \cap \mathcal{V}_2 \cap \cdots \cap \mathcal{V}_p.$$

Utilizing this notation, we can more formally define the set of natural numbers as follows:

$$\mathbb{N} := \bigcup_{n=1}^{\infty} \{n\} = \{1\} \cup \{2\} \cup \{3\} \cup \cdots = \{1, 2, 3, \cdots\}. \quad (5.18)$$

Sets expressible in this manner are called *countable sets*. These are sets that can be created through an iterative process such as $\mathbb{N}$ in (5.18). In other words, there is a one-to-one correspondence between each element in the set and the elements of the natural numbers,[†] which is to say that one can assign an integer value to every element in the set and count, one-by-one, the elements in the set.

Another example of a countable set would be the set of integers, defined by

$$\mathbb{Z} := \bigcup_{n=0}^{\infty} (\{-n\} \cup \{n\})$$
$$= (\{-0\} \cup \{0\}) \cup (\{-1\} \cup \{1\}) \cup (\{-2\} \cup \{2\}) \cup \cdots$$
$$= \{0\} \cup \{-1, 1\} \cup \{-2, 2\} \cup \cdots$$
$$= \{0, -1, 1, -2, 2, \cdots\}$$
$$= \{\cdots, -2, -1, 0, 1, 2, \cdots\}.$$

This is a countable set because there exists a one-to-one correspondence between each element in $\mathbb{Z}$ and the natural numbers $\mathbb{N}$. To

---

[†] In more mathematical terms, we would say there exists an *injection* between all countable sets and $\mathbb{N}$.

see how, just continue the pattern:

$$1 \mapsto 0$$
$$2 \mapsto 1$$
$$3 \mapsto -1$$
$$4 \mapsto 2$$
$$5 \mapsto -2$$

There is a clear one-to-one correspondence between the integers (right column) and the natural numbers (left column). Not only does it follow that $\mathbb{Z}$ is countable, but also that $|\mathbb{Z}| = |\mathbb{N}|$—that the set of integers has *the same cardinality* as the set of natural numbers.[†]

Because $\mathbb{N}$ and $\mathbb{Z}$ are both countable and of infinite cardinality,[‡] they are called *countably infinite* sets. Sets that are not countably infinite are either finite sets or *uncountably infinite* sets. Finite sets are not of huge interest to us. Uncountably infinite sets, however, are. These sets are literally larger than countably infinite sets because no one-to-one correspondence exists between the elements in the set and the natural numbers. In other words, it is not possible to count the elements in an uncountable infinite set (hence the name *uncountable*) because it is impossible to list even a portion of the set in increasing order. It follows that uncountably infinite sets have more elements than the set of natural numbers.

To illustrate, consider the set of real numbers $\mathbb{R}$. To count the elements in the set, we require the elements in the set. For sake of simplicity, we omit all negative and rational numbers and start counting at zero. But what is the next real number following zero? 0.1? 0.0001? 0.00000000001? There is no answer to this question,

---

[†] In a similar manner, it can be shown that the set of even numbers is of identical cardinality to $\mathbb{N}$, and, perhaps more astonishingly, that $|\mathbb{Q}| = |\mathbb{N}|$.

[‡] Here we would say there exists a *bijection* (a case where both an injection and something called a *surjection* are present) between $\mathbb{Z}$ and $\mathbb{N}$.

§5.4. Set Theory: $\infty > \infty$

for we can always append an additional zero to our answer. So it seems that we cannot list all the elements in $\mathbb{R}$.

A more clever demonstration of this property was uncovered by the nineteenth-twentieth century mathematician Georg Cantor. His famous *diagonal argument* proceeds as follows: Suppose $\mathbb{R}$ is countably infinite. Then, by definition, we can count all the elements in $\mathbb{R}$, which necessarily implies we can list all the elements in $\mathbb{R}$. Suppose, without loss of generality, that the following elements are the first few in $\mathbb{R}$:

$$1 \mapsto 2.46372843990843759\cdots$$
$$2 \mapsto 3.23784639439840202\cdots$$
$$3 \mapsto 4.54834950229083472\cdots$$
$$4 \mapsto 5.23904932765432895\cdots$$
$$5 \mapsto 7.87329539398734598\cdots$$

We will show that from the numbers listed, we can always construct a new number $\Psi \in \mathbb{R}$ that is not yet listed. To do this, we summon the piecewise function $\psi(n)$ defined by

$$\psi(n) := \begin{cases} 0 & \text{if } n = 9 \\ n+1 & \text{otherwise.} \end{cases}$$

We now proceed through the set of elements in $\mathbb{R}$ and formulate a new number by evaluating $\psi(n)$ on the first digit of the first number, then $\psi(n)$ on the second digit of the second number, then $\psi(n)$ on the third digit of the third number, etc. Our new number Ψ is the concatenation of all the $\psi(n)$. Using the values above, the process is thus:

$$\boxed{2}.46372843990843759\cdots \quad\quad \boxed{3}.46372843990843759\cdots$$
$$3.\boxed{2}3784639439840202\cdots \quad\quad 3.\boxed{3}3784639439840202\cdots$$
$$4.5\boxed{4}834950229083472\cdots \xrightarrow{\psi} 4.5\boxed{5}834950229083472\cdots$$
$$5.23\boxed{9}04932765432895\cdots \quad\quad 5.23\boxed{0}04932765432895\cdots$$
$$7.873\boxed{2}9539398734598\cdots \quad\quad 7.873\boxed{3}9539398734598\cdots$$

making our new number $\Psi = 3.3503\cdots$. Notice that Ψ cannot be the first number in the set of $\mathbb{R}$ because it is different in the first place. Similarly, Ψ cannot be the second number because it is different in the second place. And in general, Ψ cannot be the nth number because it is different in the nth place. Hence, we have constructed a number that was not previously listed in the set of $\mathbb{R}$. More profoundly, though, no one is stopping us from repeating the same process an arbitrary number of times, at each iteration deducing a new number not previously in $\mathbb{R}$. It follows that no matter how many elements we list there can always be more listed, meaning we cannot possibly count all the elements in $\mathbb{R}$. In other words, we cannot assign a one-to-one correspondence between the elements in $\mathbb{R}$ with those in $\mathbb{N}$. Hence, by definition, $\mathbb{R}$ must be uncountably infinite. And from our discussion on countably versus uncountably infinite sets, it follows that $|\mathbb{R}| > |\mathbb{Z}|$, suggesting that some infinities are literally larger than others.

We'll now philosophize for a moment and end on a paradoxical mathematical construction formulated by the twentieth century mathematician and philosopher Bertrand Russell. In an attempt to construct logical paradoxes within mathematics, Russell came across the idea of constructing a

set of all sets that does not contain itself.

Such a set can be difficult to envision, perhaps for good reason: A set with this property is a contradictory construction. Consider, for instance, that such a set does exist, which we call $\mathcal{Z}$. By its definition, the set $\mathcal{Z}$ contains all possible sets. But $\mathcal{Z}$ is itself a set. Hence, $\mathcal{Z}$ must contain itself. This is a contradiction because, by its definition, $\mathcal{Z}$ does not contain itself. Consequently, we must conclude that no such $\mathcal{Z}$ exists.

§5.4. Set Theory: $\infty > \infty$

Yet one can go further, and define $\mathcal{Y}$ as the

set of all sets which are not members of themselves.

This construction is more than contradictory, it is entirely paradoxical, hence its name: *Russell's paradox*. Formally, $\mathcal{Y}$ is notated as

$$\mathcal{Y} := \{\mathcal{X} : \mathcal{X} \notin \mathcal{X}\},^\dagger$$

where $\mathcal{X}$ is any set that does not contain itself. To formulate the paradox, suppose $\mathcal{Y}$ exists and assume $\mathcal{Y} \in \mathcal{Y}$. However, this assumption implies $\mathcal{Y}$ contains itself, so it should not be in the set. Hence, $\mathcal{Y} \notin \mathcal{Y}$. That is, $\mathcal{Y}$ is a set that does not contain itself. But this implies $\mathcal{Y} \in \mathcal{Y}$ by the definition of $\mathcal{Y}$. But this self-containment was just shown erroneous. As you can see, unless we put an end to the logical flow, we'll get stuck in an infinite logical loop. Altogether, this renders the notion of $\mathcal{Y}$ paradoxical, forcing the conclusion that $\mathcal{Y}$ cannot exist. Hence, there is no set $\mathcal{Y}$ such that $\mathcal{Y}$ is the set of all sets which are not members of themselves.

This conclusion may seem rather unimportant. In truth, however, it demonstrates an intrinsic paradox within our definition of a set, which we naïvely defined as a collection of distinct objects. A collection of distinct objects is, of course, quite broad, and does not rule out anyone from saying that these objects are sets with some property. Such an assignment on the objects gives rise to *Naïve Set Theory*—an informal theory of sets that presumed all sets of all sets with some property existed. But as we've just shown, if this property is that the elements of the set are sets such that they do not contain themselves, then a paradox is born. Hence, naïve set theory is a self-contradictory theory because it presumes there are sets which can be composed that do not actually exist. That is, unless more restrictions are imposed on the possible sets constructible in naïve

---

†This is read "$\mathcal{Y}$ is the set of all $\mathcal{X}$ such that $\mathcal{X} \notin \mathcal{X}$."

set theory, which is what many mathematicians have done (this includes abandoning naïve set theory and devising other theories of sets, such as *Zermelo set theory*, the progenitor for our modern theory of sets, after the nineteenth-twentieth century mathematician Ernst Zermelo).

§5.5 Multivariate Calculus: A Hole in the Earth

Multivariate calculus is the extension of single-variable calculus to higher dimensions. Rather than computing slopes and areas beneath two-dimensional lines, multivariate calculus is concerned with the "slopes" of surfaces (called the *gradient*, denoted ∇), volumes under surfaces, integrals along paths, etc. Indeed, concepts become more complex and more difficult to visualize, but the mathematics is similar and always enticing.

Multivariate calculus is also the language through which some of society's greatest achievements were first communicated. For instance, it is the language of electromagnetic theory in physics, which accounts for the function of all the electronics we take for granted on a daily basis, such as our mobile phones. That said, we do not concern ourselves with electromagnetism here. Instead, we explore a union of multivariate calculus and mechanical physics to compute the time it would take to fall through a hole in the Earth.

In §3.3, we introduced Newton's law of gravity

$$F = -\frac{GMm}{r^2}, \qquad (5.19)$$

where F is the gravitational force acting on the object of mass m by mass M, r is the radial distance from the center of mass M to the center of mass m, and G is the gravitational constant. If R_M is the radius of the spherical mass M, then (5.19) only applies for $r \geq R_M$. Here, however, we will be jumping (or rather falling) through the Earth. This naturally requires an understanding of the gravitational

§5.5. Multivariate Calculus: A Hole in the Earth

force F *inside* the spherical mass M—that is, for $r < R_M$. As you can guess, this necessitates an extension into multivariate calculus.

Though before diving in, a word on what follows. We have omitted a majority of the extensive and distinct notation of multivariate calculus so that our analysis is not obfuscated. Moreover, we perform this analysis without mathematical objects/operations like vectors, gradients, divergence, and so forth, and instead settle for a more qualitative approach wherein we use language over symbols.

Our solution begins by acknowledging one of the most fundamental of theorems in all multivariate calculus: *Gauss' theorem* (i.e. the *divergence theorem*). This theorem applies to three-dimensional objects and their surfaces, such as an apple (the object) and the apple's skin (its surface). Picture a sphere filled with water that has many small holes throughout its surface. Of course, the water will be leaking out, but this is what we want. Gauss' theorem tells us that the total amount of water flowing out of the surface of the sphere is proportional to the amount of water flowing throughout the volume of the sphere enclosed by the holed surface. In other words, there is a link between the flow of water on the surface of the sphere to the flow of water on the inside of the sphere.

Now apply the analogy to gravity, where instead of water we have an emanating gravitational field from a spherical planet of mass M. The strength of the gravitational field is determined by (5.19) above. Gauss' theorem tells us that the flow of the gravitational field (this is called the *flux* of the gravitational field) at the surface of the planet is proportional to the flow (i.e. flux) throughout the volume of the sphere that is enclosed by the surface.

To construct this symbolically, first recall Newton's second law $F = ma$. Combining this with (5.19), it follows that

$$a = g(r) = -\frac{GM}{r^2},$$

where $g(r)$ is the gravitational acceleration a distance r away from the center of mass M. It is productive to think of this function as one which assigns a value to all points in space around the planet of mass M. While this is a gravitational field, it is, again, for $r \geq R_M$. Hence, we wish to find the gravitational field $g(r)$ for $r < R_M$. To do this, we apply Gauss' theorem inside the spherical planet by constructing a concentric sphere with radius $r < R_M$ inside the planet. Indeed, Gauss' theorem applies independently for this new sphere. This means that all gravitational effects from any concentric, spherical shell outside the new sphere need not be considered. Though counterintuitive, this is how nature operates.

The theorem tells us the flow of the gravitational field at the surface of our new sphere, which is simply a measure of the strength of the field at the surface of the sphere, is proportional to the flow throughout the volume of this new sphere, which, again, is a measure of the strength of the field by the whole sphere. In other words, the strength of the gravitational field at the surface of the new sphere is proportional to the strength of the field caused by the spherical ball enclosed by this new surface. But this is almost equivalent to what (5.19) says about any sphere. And indeed it is the case that for $r < R_M$,

$$F = -\frac{GM_{\text{enc}}m}{r^2} \implies g(r) = -\frac{GM_{\text{enc}}}{r^2}, \qquad (5.20)$$

where M_{enc} is the mass of the new, concentric, and enclosed sphere. Consequently, all we require to deduce $g(r)$ for $r < R_M$ is an expression for M_{enc}.

Assuming the mass M is a spherical planet of constant density $\rho = \frac{M}{V}$, where V is the volume of the mass M, it follows that

$$M_{\text{enc}} = \rho V_{\text{enc}},$$

where V_{enc} is the volume of the enclosed sphere. For any sphere of radius R, we know the volume to be

$$V = \frac{4}{3}\pi R^3.$$

§5.5. MULTIVARIATE CALCULUS: A HOLE IN THE EARTH

This makes the density of the spherical planet

$$\rho = \frac{M}{V} = \frac{3M}{4\pi R^3}.$$

Hence, if r is the radius of the enclosed sphere, $V_{\text{enc}} = \frac{4}{3}\pi r^3$ and so

$$M_{\text{enc}} = \underbrace{\left(\frac{3M}{4\pi R^3}\right)}_{\text{density } \rho}\left(\frac{4\pi r^3}{3}\right) = \frac{Mr^3}{R^3}.$$

Plugging into (5.20), we obtain

$$g(r) = -\frac{GM}{R^3}r,$$

which is to say that for $r < R$, gravity decreases linearly in the radius r as one approaches the center of the planet. We now have sufficient ground to approximate the time to fall through a hole in the Earth.

Because gravity is a form of acceleration and, in turn, acceleration is the second derivative of position (encountered in §3.3), the governing differential equation is

$$a = g(r) \implies \frac{d^2r}{dt^2} = -\frac{GM}{R^3}r. \qquad (5.21)$$

We prove in the appendix (§A.9) that the general solution to such a differential equation is

$$r(t) = A\cos(\omega t) + B\sin(\omega t),^\dagger$$

where

$$\omega = \sqrt{\frac{GM}{R^3}}$$

is the *angular frequency* (i.e. periodicity) of oscillation and A, B are constants of integration.[‡] Why is there oscillation? Simply because there are trigonometric functions present.

[†] The reader is encouraged to verify that this does indeed satisfy the differential equation above. Simply differentiate it twice to do so.

[‡] (5.21) is a *second-order* differential equation, so it's integrated twice to obtain r, introducing two constants of integration.

Given the angular frequency, the period T for a single oscillation is given by
$$T = \frac{2\pi}{\omega},$$
which is to say
$$T = 2\pi \sqrt{\frac{R^3}{GM}}.$$
Inputting Earth's astronomical data ($R \approx 6.37 \times 10^6$ m and $M \approx 5.97 \times 10^{24}$ kg), we find the time for someone to go through an oscillation inside the Earth to be approximately

$$T \approx 5062.2 \text{ seconds}.$$

In more reasonable units, this is about 84.4 minutes per oscillation. But we are concerned with just going from one side of the Earth to the other, not there are back. We simply divide the 84.4 minute result by two to obtain

42.2 minutes—the time to fall through a hole in the Earth.

By the trigonometric nature in the solution to the above differential equation, the person jumping in the hole will oscillate between antipodal sides of the Earth forever (all other forces absent, such as drag). Of course, in reality drag forces, among others, would be present, which would slow the periodicity of the oscillation as well as decrease the amplitude of each cycle. This means that once you jump into the hole, you may never be able to get back out. That said, the Earth's geothermal furnace would be the greatest concern. It turns out humans do not work well in temperatures around $10,400°F$—the temperature of the Earth's inner core.

§5.6 Mathematical Physics: Relativity

Albert Einstein is famous for many, many reasons. His scientific contributions span from the very small of quantum mechanics to the

§5.6. MATHEMATICAL PHYSICS: RELATIVITY

very large of theoretical cosmology. Ask anyone about Einstein and they will instantaneously recite his famous equation

$$E = mc^2.^\dagger$$

Though this equation, believe it or not, was postscript in his formulation of what is coined the *Special Theory of Relativity*—the physical theory concerned with the counterintuitive responses of both space and time when the velocity of particles are very near the speed of light. Einstein's fame, however, is primarily derived from his formulation of the *General Theory of Relativity* and his contributions to something called the *Photoelectric Effect*,[‡] the latter for which he was awarded the Nobel Prize in 1921. In this section we explore two astounding consequences of the special and general theories of relativity (together these theories are simply called *relativity*): time dilation and black holes.

Time dilation is the first major result in relativity. It maintains that time, the continuum of events, is not absolute—that the very passage of time is dependent upon one's frame of reference. In other words, there is no universal clock ticking away the seconds exactly at the rate as every other clock. What you measure as ten seconds we may very well measure as a hundred.

[†] Here, E is energy, m is mass, and $c = 299,792,458$ ms^{-1} is the speed of light. Though famous, this equation is actually a special case of the more general equation $E^2 = (mc^2)^2 + (pc)^2$, where p is the momentum for a particle of mass m. $E = mc^2$ is derived when $p = 0$—that is, when the particle is stationary. Notice that with $m = 0$ the formula reduces to $E = pc$. Because light is known to have energy (you feel the heat of the Sun), it follows that massless photons (light particles) will still have momentum $p = \frac{E}{c}$.

[‡] General relativity is the modern theory of gravity that superseded Newton's formulation (see §3.13) in 1915. It describes the interaction between mass and spacetime (the marriage of space and time). On the other hand, the photoelectric effect explains the mechanics behind the emission of electrons (the negatively charged particles that orbit atomic nuclei) when photons bombard a material.

Such a notion on the passage of time *not* being absolute should be counterintuitive. If not, then you have yet to truly indulge in what we are talking about. We will soon illustrate this phenomenon conceptually, and from this derive the necessary mathematics. Before this, however, we require a small introduction to the postulates of relativity. And what better place to start than where Einstein himself did: a thought experiment.

Imagine yourself riding through space in a rocket. Convince yourself that if moving at a perfectly constant velocity (say, with your eyes closed), there is no way to tell whether you are moving or are stationary (again, absent all accelerations). In an airplane, for instance, gravity remains downward, the cries of the baby still emanate throughout the fuselage, etc. The point is, the laws of physics do not change whether you are moving at a constant speed or at no speed. Absent all acceleration, a frame of reference in which you are subject to either constant or no speed is called an *inertial frame of reference* (terminology we encountered in §3.3). This brings us to the first postulate of relativity:

> *The laws of physics are the same in all inertial reference frames.*

In 1865, the great physicist James Clerk Maxwell demonstrated that the speed of light is independent of one's reference frame by deriving the speed of light irrespective to a particular vantage point. Though this may seem to be of little consequence, it is, at the very least, counterintuitive.

Consider, for instance, us traveling past you in a car moving at a constant 10 ms^{-1} in which we throw a ball in our direction of travel at 5 ms^{-1} relative to us. Of course, you would see the ball move at 15 ms^{-1}. But if the ball instead moves at the speed of light relative to us, Maxwell unknowingly proved that you would not measure the ball's speed as the speed of light plus 10 ms^{-1}, but rather at the same speed as us—the speed of light. And this is true regardless of

§5.6. MATHEMATICAL PHYSICS: RELATIVITY

how fast our car is moving relative to you. In other words, despite one's frame of reference, the speed of light is always constant. This leads us to the second and last postulate of relativity:

> *The speed with which light travels is constant in all inertial reference frames.*

We'll now get more mathematical. Suppose you, back in the spaceship, travel past an identical spaceship with speed v. At the moment the two ships are exactly next to one another, you shine a laser from the floor of your spacecraft to the ceiling and time the beam from the bottom to the top. In your reference frame, this beam travels exactly vertically. If the height of your ship is h and the speed of light c, then the time interval $\Delta \tau$ it takes the light to travel from the bottom to the top of your ship is simply

$$\Delta \tau = \frac{h}{c}. \tag{5.22}$$

Now consider the same experiment from the perspective of the ship you pass by. Relative to them, the point at which you initially shine the laser is ahead of the point at which the laser strikes the ceiling since, to them, your ship is moving by at speed v. That is, rather than traveling vertically (as the beam did for you), the beam travels diagonally relative to the observers on the other ship. It follows that the beam traverses a longer distance relative to them. But because the speed of light is the same in every reference frame (this is the second postulate of relativity), it must be that the time it took for the light to reach the ceiling in your spaceship was shorter than for the outside observing spaceship.

To derive the precise temporal relationship, first let Δt be the time interval as measured by the observers on the other spacecraft. Relative to them, the light traveled up a distance h but also back a distance $v\Delta t$, because, relative to them, your spaceship is moving forward at speed v. Hence, the total distance d traveled by the light beam is obtained using the Pythagorean theorem:

$$d = \sqrt{v^2 \Delta t^2 + h^2}.$$

Therefore the total time interval Δt is

$$\Delta t = \frac{d}{c} = \frac{1}{c}\sqrt{v^2 \Delta t^2 + h^2} \implies \Delta t^2 = \frac{v^2}{c^2}\Delta t^2 + \frac{h^2}{c^2}.$$

But the fraction $\frac{h^2}{c^2}$ is the square of (5.22). Thus,

$$\Delta t^2 = \frac{v^2}{c^2}\Delta t^2 + \Delta \tau^2$$

which simplifies to

$$\Delta t = \frac{\Delta \tau}{\sqrt{1 - \frac{v^2}{c^2}}}. \tag{5.23}$$

This equation is the famous *time dilation* formula and is often expressed more succinctly as

$$\Delta t = \gamma \Delta \tau$$

where

$$\gamma := \frac{1}{\sqrt{1 - \left(\frac{v}{c}\right)^2}}$$

is called the *Lorentz factor* after the esteemed nineteenth-twentieth century physicist Hendrik Lorentz.

In its most profound sense we have just shown that time is not absolute—that time does not tick by at the same rate for everyone. More concretely, we have shown that for an event in your reference frame which, to you, took a time interval $\Delta \tau$ to occur, observers traveling past you at speed v will time the same event as being the longer $\gamma \Delta \tau$. More generally, moving clocks run slower than stationary ones.

But let's not stop here. Notice in the limit as $\Delta \to 0$ we can write (5.23) in differential form as

$$dt = \gamma d\tau. \tag{5.24}$$

§5.6. Mathematical Physics: Relativity

Recall from §3.3 that velocity is the first time derivative of position x. That is,
$$v_x = \frac{dx}{dt}.$$
This expression, however, only accounts for velocity in one dimension along the x-axis. The total velocity in three-dimensional space is the Pythagorean-like sum
$$v = \sqrt{v_x^2 + v_y^2 + v_z^2} = \sqrt{\left(\frac{dx}{dt}\right)^2 + \left(\frac{dy}{dt}\right)^2 + \left(\frac{dz}{dt}\right)^2}.$$
Substituting this velocity into (5.24) we obtain
$$dt = \frac{d\tau}{\sqrt{1 - \frac{\left(\frac{dx}{dt}\right)^2 + \left(\frac{dy}{dt}\right)^2 + \left(\frac{dz}{dt}\right)^2}{c^2}}}.$$
This formulation is easily manipulated if we allow ourselves to treat the differentials as fractions. Upon doing so, we can square both sides and solve for $c^2 d\tau^2$:
$$c^2 d\tau^2 = c^2 dt^2 - dx^2 - dy^2 - dz^2. \tag{5.25}$$

This conglomerate of squared differentials is called a *line element* and is quite a ubiquitous expression in the field of *differential geometry*. The next few paragraphs are devoted to interpreting it.

The meaning of (5.25) is best understood by considering the simpler line element
$$ds^2 = dx^2 + dy^2. \tag{5.26}$$
Here ds is a distance—an infinitesimally small interval of distance, to be precise—oriented some way in the x, y-plane, like that in Figure 5.1. Because ds exists in two-dimensional Euclidean space (the x, y-plane), it has some x component dx and some y component dy. We can therefore describe ds in terms of dx and dy by using the Pythagorean theorem: $ds^2 = dx^2 + dy^2$, which is (5.26).

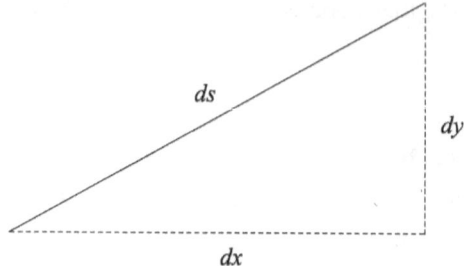

Figure 5.1: The line element $ds^2 = dx^2 + dy^2$.

From this example, it seems that the line element details the relationship between small distances in space and the coordinates of that space. While this is true, there is more to it than that. Imagine "adding up" (by means of an integral) all these infinitesimal distances described in the line element, over all possible permutations of the coordinates. The result would be all possible paths s in that space. Concatenating all such paths would then construct the space itself. Hence, the line element tells us about the structure of the space, which is to say it encodes information about the *geometry of the space*.

Fundamentally, then, (5.25) tells us about the geometry of some space. The exact details of this space are beyond our discussion here, but we do mention that the space for which (5.25) applies contains three spatial dimensions (e.g. x, y, and z) and a temporal one (namely t). Therefore, the space is four dimensional, composed of both spatial and temporal components. In the relevant literature, this union of space and time is called *spacetime*—it is the underlying geometry of the universe.

Einstein's great achievement in his general relativity was deducing that spacetime geometry is influenced by massive objects, and, in turn, that massive objects respond to changes in spacetime

§5.6. MATHEMATICAL PHYSICS: RELATIVITY

by moving. We include the corresponding equation for marveling purposes:

$$R_{\mu\nu} - \frac{1}{2}Rg_{\mu\nu} + \Lambda g_{\mu\nu} = \frac{8\pi G}{c^4}T_{\mu\nu}. \tag{5.27}$$

In theoretical physics, this equation is considerably more important than $E = mc^2$. To describe (5.27) in layman's terms, we turn to the famous words of renowned physicist John Archibald Wheeler, "Matter tells space how to curve. Space tells matter how to move."

The line element in (5.25) describes a special type of spacetime. It is absent massive objects such as stars and planets, and so is sometimes described as *flat spacetime* or *Minkowski space* (after the twentieth century mathematician Hermann Minkowski) because there are no objects nearby to warp it any which way. That said, a spacetime line element need not depict a flat geometry. The curvature of spacetime can become very large near supremely dense objects such as *black holes*—celestial bodies whose gravitational attraction not even light can escape. In more mathematical terms, a black hole is an object whose density is so great that the local curvature of spacetime has gone infinite.

We illustrate some bizarre consequences of black holes by examining the line element near a spherical, non-rotating black hole[†] whose local geometry is anything but flat:

$$c^2 d\tau^2 = c^2\left(1 - \frac{R_s}{r}\right)dt^2 - \frac{dr^2}{\left(1 - \frac{R_s}{r}\right)} - r^2 d\theta^2 - r^2 \sin^2(\theta)d\phi^2. \tag{5.28}$$

Here

$$R_s := \frac{2GM}{c^2}$$

is the *Schwarzschild radius* of the black hole, named after the twentieth century astronomer Karl Schwarzschild. His radius is that at which a spherical object of mass M will collapse in on itself to form

---

[†] A black hole with these properties is called a *Schwarzschild black hole*.

a black hole.[†] The Schwarzschild radius is also the distance from the black hole's center (called its *singularity*) to something called its *event horizon*. In layman's terms, the event horizon is the point of no return near a black hole. Journey past it even slightly and nothing, including light, can escape the black hole's gravitational attraction. Inevitably, passing the event horizon guarantees you being drawn in and meeting your demise.

Going back to the line element in (5.28), we see some other symbols not present in the simpler (5.25). Ignoring the $d\theta$ and $d\phi$ (these concern the fact that the local spacetime geometry is approximately spherical), the r represents an observer's radial distance from the singularity. Relating to our discussion on the event horizon, for $r > R_s$ it is still possible to get away from the black hole and survive. However, for $r < R_s$ (past the event horizon) neither you nor light can escape the gravitational attraction.

For the particular case where $r = R_s$ something interesting happens. To properly illustrate the effect, we divide each side of (5.28) by dt^2. This yields

$$c^2 \left(\frac{d\tau}{dt}\right)^2 = c^2 \left(1 - \frac{R_s}{r}\right) - \left(1 - \frac{R_s}{r}\right)^{-1} \left(\frac{dr}{dt}\right)^2$$
$$- r^2 \left(\frac{d\theta}{dt}\right)^2 - r^2 \sin^2(\theta) \left(\frac{d\phi}{dt}\right)^2.$$

Here, we can think of $d\tau$ as an infinitesimal amount of time for someone falling into the black hole, and dt as an infinitesimal amount of time for someone outside the black hole who is watching the other person fall in. The ratio $\frac{d\tau}{dt}$ is therefore the rate at which time passes for the infalling observer compared to the outside observer. For simplicity we assume the observer falls into the black hole radially inward, so there is no rotational movement. This means the

---

[†]The Schwarzschild radius of the Earth, for instance, is about 0.9 centimeters— roughly the size of a peanut M&M. This means that if the Earth were crushed down to such a size, it would gravitationally collapse in on itself and form a black hole.

§5.6. Mathematical Physics: Relativity

quantities governing the rate of rotation—namely, $\frac{d\theta}{dt}$ and $\frac{d\phi}{dt}$—are zero. Hence, the above formulation simplifies to

$$c^2 \left(\frac{d\tau}{dt}\right)^2 = c^2 \left(1 - \frac{R_s}{r}\right) - \left(1 - \frac{R_s}{r}\right)^{-1} \left(\frac{dr}{dt}\right)^2. \quad (5.29)$$

As you, the infalling observer, approach the event horizon (by this we mean $r \to R_s^+$, where the + means from larger to smaller r—so approaching from outside to inside the black hole), (5.29) becomes

$$c^2 \left(\frac{d\tau}{dt}\right)^2 = \lim_{r \to R_s^+} \left[c^2 \left(1 - \frac{R_s}{r}\right) - \left(1 - \frac{R_s}{r}\right)^{-1} \left(\frac{dr}{dt}\right)^2 \right]$$
$$\to \infty.$$

That is, $\frac{d\tau}{dt} \to \infty$ as $r \to R_s^+$. The physical meaning of such a limit is absolutely staggering: Relative to a distant observer watching you approach the event horizon ($r \to R_s^+$), infinitesimal changes in time for them take an infinite amount of time for you. Physically, relative to the the outside observer, as you approach the event horizon your motion in time slows down to the point where you appear to freeze in space and time. To them, you essentially become glued to the side of the black hole while your figure fades into transparency as the light reflected off your body "burns up" due to gravitational effects.[†]

Naturally, for you the infalling observer, the story is quite different.[‡] Depending on the size of the black hole, the difference in gravity between your feet and head (a form of something called a *tidal force*) would be small both outside and inside the event horizon, so you may not even notice when you've passed this point of no return. But as you proceed down into the darkness of the black hole, the tidal forces—the differences in gravity across your body—will

---

[†] By this we mean the light becomes infinitely *redshifted*—that is, the wavelength of the light reflected off your body increases to infinity (rendering it impossible to see) due to the immense gravity of the black hole.

[‡] There are many different theories for what the infalling observer would witness. We provide the most widespread version here.

begin to squeeze and stretch you, deforming you into an infinitesimally thin piece of wire—a process called *spaghettification*. This would likely be a painful departure.

Besides the conclusion that mathematical and theoretical physics are enticing applications of mathematics, the moral of this story is simple: If journeying through interstellar space, it's mildly irrational to play around black holes. As we have seen, they are quite an unforgiving celestial species.

§5.7 Research Mathematics: Open Conjectures

Beyond all the topics and ideas discussed so far, none are as compelling and invigorating as the forefront of mathematical research. It was here where all the topics discussed in this book were originally devised in the abstract, and later found to have revolutionary implications. Indeed, many of our discussions were originally posed as conjectures, and their truth assumed well before their proof. A modern day example of such an open conjecture whose proof is assumed true is the esteemed Riemann hypothesis. Perhaps you will be the one to uncover its proof or, perchance, discover the first counterexample.

Though it is wrong of us to point you in the sole direction of Riemann. Instead, after our pursuit into some known problems and their amazing implications, it seems only right to end with a handful of additional, contemporary conjectures, each whose proof would grant you eternal fame in the world of mathematics. The following conjectures are listed in increasing complexity, but that does not necessarily mean in increasing difficulty. We also restate a few conjectures found earlier in the text.

The Collatz Conjecture

The *Collatz conjecture* (also known as the $3n + 1$ Conjecture), named after the mathematician Lothar Collatz, concerns the recursive and piecewise function

$$C(n) := \begin{cases} 1 & \text{if } n = 1 \\ C(\tfrac{n}{2}) & \text{if } n \text{ is even} \\ C(3n+1) & \text{if } n \text{ is odd,} \end{cases} \quad (5.30)$$

where n is a positive integer. The conjecture is:

Conjecture 5.1 (Collatz Conjecture). *$C(n) = 1$ for all positive integers n.*

To illustrate $C(n)$, consider the case with $n = 5$. Since five is odd, it follows from (5.30) that

$$C(5) = C(3 \times 5 + 1) = C(16).$$

Sixteen is even, so

$$C(16) = C\left(\frac{16}{2}\right) = C(8),$$

which is even, making

$$C(8) = C\left(\frac{8}{2}\right) = \underbrace{C(4)}_{\text{even } n} = \underbrace{C(2)}_{\text{even } n} = C(1).$$

But $C(1) = 1$. Ergo,

$$C(5) = 1,$$

as conjectured.

Do indulge yourself with larger numbers. It is quite satisfying seeing them all trickle down to unity. To simplify the computation, take note of the recursive structure of the conjecture. This begs to be implemented by a computer program. For those interested, our Python transcript begins on page 261.

The Twin Prime Conjecture

The *twin prime conjecture* studies pairs of prime numbers called *twin primes*. If p_n denotes the nth prime, then a twin prime is a pair p_n, p_{n+1} such that
$$p_{n+1} - p_n = 2.$$
The conjecture is:

Conjecture 5.2 (Twin Prime Conjecture). *There exist infinitely many twin primes.*

The first seven twins primes are:

$(3,5), (5,7), (11,13), (17,19), (29,31), (41,43),$ and, $(59,61),$

which were generated from the algorithm on page 261.

The Goldbach Conjecture

Stately simply, the *Goldbach conjecture* concerns the prime numbers and addition. The conjecture is:

Conjecture 5.3 (Goldbach Conjecture). *Every even integer greater than two can be expressed as the sum of two primes.*

To demonstrate, consider the integers: $42, 1120,$ and 101010. Our Goldbach computer algorithm (page 261) outputs

$$42 = 5 + 37$$
$$1120 = 3 + 1117$$
$$101010 = 11 + 100999.$$

Because $3, 5, 11, 37, 1117,$ and 100999 are all prime, these sums verify the conjecture for $42, 1120,$ and 101010.

§5.7. Research Mathematics: Open Conjectures

The Euler-Mascheroni Constant

We first encountered the Euler-Mascheroni constant in §4.3. As a reminder, the Euler-Mascheroni constant, denoted γ, is defined as

$$\gamma := \lim_{m \to \infty} H_m - \log(m) = 0.577215 \cdots,$$

where

$$H_m := \sum_{n=1}^{m} \frac{1}{n}$$

is the mth harmonic number.

The conjecture asks:

Conjecture 5.4. *Is the Euler-Mascheroni constant γ irrational?*

Rather than provide some sort of example, we find it pleasing to list a larger decimal expansion of γ while contemplating the conjecture:

$$\gamma = 0.57721566490153286060651209008240243104215933\cdots.$$

Perfect Numbers

Perfect numbers are numbers whose divisors add up to twice the number. In other words, if $d_1, d_2, \cdots, d_k$ are the divisors of the positive integer n, and if the sum

$$d_1 + d_2 + \cdots + d_k = 2n,$$

then n is a perfect number. Mathematicians prefer to condense this notation by defining a function called the *sum-of-divisors function*, denoted $\sigma(n)$, whose name tells the whole story:

$$\sigma(n) := \sum_{i=1}^{k} d_i,$$

where, as above, $d_1, d_2, \cdots, d_k$ are the divisors of n. Using this notation, a perfect number is one for which $\sigma(n) = 2n$. Some authors will condense this notation further by writing

$$\sigma(n) := \sum_{d|n} d.$$

The $d|n$ we encountered in §4.5 and reads "d divides n." Therefore, the sum above is over all d such that d divides n—precisely the sum over all divisors of n.

To illustrate, consider the case with $n = 6$. The divisors of six are $1, 2, 3,$ and 6. Adding these values we obtain:

$$1 + 2 + 3 + 6 = 12 = 2 \times 6,$$

making six a perfect number. Our algorithm on page 261 lists the first five perfect numbers as

$$6, 28, 496, 8128, \text{ and } 33550336,$$

all of which are even.

The two primary conjectures concerned with perfect numbers ask:

Conjecture 5.5. *Do any odd perfect numbers exist?*

and

Conjecture 5.6. *Are there infinitely many even perfect numbers?*

A proof of the latter would immediately prove a related conjecture on the infinitude of *Mersenne primes*, which are primes p of the form

$$p = 2^m - 1,[†]$$

where $m \in \mathbb{N}$. Refer to §A.10 for a proof on the direct correspondence between Mersenne primes and perfect numbers.

[†] Mersenne primes are named after the seventeenth century theologian, philosopher, and mathematician Marin Mersenne.

Lehmer's Totient Problem

Lehmer's totient problem is named after the twentieth century mathematician Derrick Lehmer. As the title of the conjecture suggests an understanding of the word "totient" is required. This, in turn, requires an understanding of coprime numbers.

Utilizing notation from number theory, we say that two numbers i and j are *coprime*, denoted $i \perp j$, if i and j share no common factors (excluding the integer one). The *totient* of the number n, denoted $\phi(n)$, is then defined as the number of positive integers less than or equal to n that are coprime with n. In other words, if w is the totient of n, then there are w positive integers $m_1, m_2, \cdots, m_w$ all less than n such that

$$m_1 \perp n, m_2 \perp n, \cdots, m_w \perp n.$$

Euler discovered that the totient of a number n is given exactly by the formula

$$\phi(n) = n \prod_{p \mid n} \left(1 - \frac{1}{p}\right), \quad (5.31)$$

where p is a distinct prime factor of n. (5.31) is called *Euler's totient function* and can be expressed in less ambiguous notation as follows: If

$$n = p_1^{\alpha_1} \times p_2^{\alpha_2} \times \cdots \times p_v^{\alpha_v}$$

is the prime factorization of n, where all $p_1, p_2, \cdots p_v$ are distinct primes and $\alpha_1, \alpha_2, \cdots \alpha_v$ are positive integers, then the totient of n is given by

$$\phi(n) = n \times \left(1 - \frac{1}{p_1}\right) \times \left(1 - \frac{1}{p_2}\right) \times \cdots \times \left(1 - \frac{1}{p_v}\right).$$

The conjecture asks:

Conjecture 5.7 (Lehmer's Totient Problem). *Are there n such that $\phi(n) = k(n-1)$ for $k > 1$?*

Perhaps not immediately obvious, we can find infinitely many examples for the case $k = 1$. Just suppose n is prime and it follows instantly from (5.31) that

$$\phi(n) = n\left(1 - \frac{1}{n}\right) = n\left(\frac{n-1}{n}\right) = n - 1.$$

Of course, the conjecture is not interested in $k = 1$ but is for $k > 1$.

The Riemann Hypothesis

We first encountered the Riemann Hypothesis in §4.11 where we were introduced to the Riemann zeta function

$$\zeta(s) := \sum_{n=1}^{\infty} \frac{1}{n^s} = \prod_{p \in \mathbb{P}} \left(\frac{p^s}{p^s - 1}\right). \qquad (5.32)$$

Here, the product on the right is over all the prime numbers in $\mathbb{P}$. In §4.11 we acknowledged the fact that (5.32) is only defined for $s \in \mathbb{C}$ such that $\Re(s) > 0$.[†] Beneath this, $\zeta(s)$ becomes

$$\zeta(s) := 2^s \pi^{s-1} \sin\left(\frac{\pi s}{2}\right) \Gamma(1-s) \zeta(1-s).$$

In total, the Riemann zeta function is really the piecewise expression

$$\zeta(s) := \begin{cases} \sum_{n=1}^{\infty} \frac{1}{n^s} & \text{if } \Re(s) > 0 \\ 2^s \pi^{s-1} \sin\left(\frac{\pi s}{2}\right) \Gamma(1-s) \zeta(1-s) & \text{otherwise.} \end{cases}$$

The Riemann hypothesis is concerned with the complex values of s such that $\zeta(s) = 0$. Recall from §4.11 that for all s of the form $s = -2k$ such that $k \in \mathbb{N}$,

$$\zeta(-2k) = 2^{-2k} \pi^{-2k-1} \sin(-k\pi) \Gamma(1+2k) \zeta(1+2k) = 0$$

[†]Recall that $\Re(s)$ is the real part of the complex number s. In other words, it is the value of a in $s = a + bi$, where $i := \sqrt{-1}$ is the complex unit.

§5.7. Research Mathematics: Open Conjectures

since the sine of any integer multiple of π is just zero. These zeros are called the *trivial zeros* of $\zeta(s)$. The Riemann hypothesis is concerned with the *nontrivial zeros* of the zeta function—that is, values of s such that $s \neq -2k$.

The conjecture is:

Conjecture 5.8 (Riemann Hypothesis). *All nontrivial zeros of the zeta function have real part $\frac{1}{2}$.*

Unlike the other conjectures in this section, the Riemann hypothesis has a one-million dollar bounty associated with it, offered by the Clay Mathematics Institute. This is one of their seven Millennium Prize Problems and is possibly the most important conjecture in mathematics.

Appendix

"Mathematics is the language of the universe. So the more equations you know, the more you can converse with the universe."
　∼ Neil DeGrasse Tyson

§A.1　On the Irrationality of $\sqrt{2}$

We proved near the end of chapter one that the nth root of two is irrational for $n > 2$. We made use of Fermat's Last Theorem to arrive at this conclusion. However, Fermat's Last Theorem will not help us prove the case where $n = 2$ because there *do* exist integer solutions to the equation $a^2 + b^2 = c^2$ (such as $3^2 + 4^2 = 5^2$). Thus, no contradiction can be fetched like that in the first chapter. Here, we resort to a more elementary (but elegant) proof for the $n = 2$ case.

Theorem A.1. *The square root of two is irrational. That is, $\sqrt{2} \notin \mathbb{Q}$.*

Proof. By way of contradiction, suppose $\sqrt{2}$ is rational. Then by the definition of a rational number, there exist integers a and b with $b \neq 0$ such that
$$\sqrt{2} = \frac{a}{b},$$
where, we assume, the fraction $\frac{a}{b}$ is irreducible (that is, a and b share no common factors). Squaring both sides, we obtain
$$2 = \frac{a^2}{b^2} \implies a^2 = 2b^2,$$

from which we deduce that the quantity a^2 is an even integer. And because a^2 is even, a must also be even (why could it not be odd?). We claim that any even number m can be expressed in the form $m = 2k$, where $k \in \mathbb{Z}$.[†] Thus, because a is even there exists a k such that

$$a = 2k \implies a^2 = 4k^2,$$

from which it follows that

$$b^2 = 2k^2.$$

By an identical argument above, we conclude that b must also be even. This, however, is a contradiction because it implies a and b have a common factor of two when we assumed they had no common factors. Consequently, $\sqrt{2}$ is irrational. $\square$

§A.2 A Note on Geometric Series

Many times throughout this book we maintain that series of the form

$$\sum_{n=0}^{\infty} C\omega^n = C + C\omega + C\omega^2 + C\omega^3 + \cdots \quad \text{(A.1)}$$

converge on the condition $|\omega| < 1$. To prove this claim, first consider the finite version of (A.1), which we denote S:

$$S := \sum_{n=0}^{k} C\omega^n = C + C\omega + C\omega^2 + C\omega^3 + \cdots + C\omega^k. \quad \text{(A.2)}$$

Take ωS and (A.2) becomes

$$\omega S = C\omega + C\omega^2 + C\omega^3 + C\omega^4 + \cdots + C\omega^{k+1}.$$

It follows that

$$S - \omega S = C - C\omega^{k+1} \implies S = \frac{C(1 - \omega^{k+1})}{1 - \omega}. \quad \text{(A.3)}$$

[†]Likewise, any odd number n can be expressed in the form $n = 2k + 1$.

We want S to remain finite as $k \to \infty$. We observe that this is only possible provided the term ω^{k+1} is finite. And the only way for this to occur is if $|\omega| < 1$, as this implies $\omega^{k+1} \to 0$ while $k \to \infty$. Hence, under the restriction $|\omega| < 1$, it follows from (A.3) that as $k \to \infty$,

$$S \to \frac{C}{1-\omega},$$

which is to say that the infinite series

$$\sum_{n=0}^{\infty} C\omega^n = \frac{C}{1-\omega}$$

provided $|\omega| < 1$.

§A.3 Derivation of Newton's Law of Gravity

In some sense, Newton's law of gravity has put the universe into our hands. It is an expression that describes the orbit of the Earth about the Sun, and the Sun about the center of the Milky Way. Its derivation for a solid sphere (approximately the Earth) is given here.

In §3.3, we assert that for two point masses m_1 and m_2 separated by a distance r, the gravitational force F on mass m_2 by m_1 is governed by the expression

$$F = -\frac{Gm_1m_2}{r^2}, \tag{A.4}$$

where $G = 6.67 \times 10^{11}$ m^{3}kg^{-1}s^{-2} is the gravitational constant. Recall that (A.4) is negative because gravity is an attractive force. To show that (A.4) generalizes beyond point masses to a spherical shell of mass m_1 and point mass of m_2, and by extension to two solid spheres like the Earth and the Moon, we consider Figure A.1. Here, R is the radius of the spherical mass m_1 (e.g. the radius of the Earth), d is the distance from the point mass m_2 to the center of m_1, and ds is an infinitesimal slice of the sphere, whose perimeter is a distance u away from m_2. For now, we assume the spherical mass

§A.3. Derivation of Newton's Law of Gravity

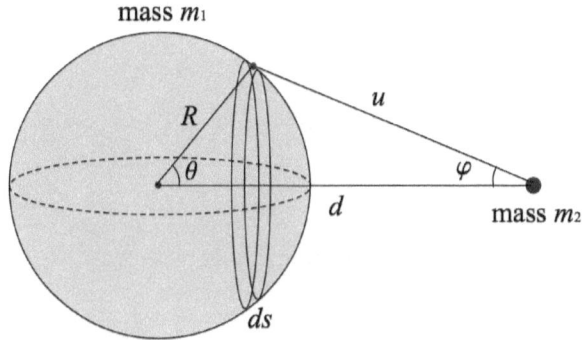

Figure A.1: A spherical shell (thickness ϵ) of mass m_1 and radius R a distance d from a point mass m_2. Here u is the distance from m_2 to an infinitesimal slice of the sphere with width ds.

is a shell of thickness ϵ. We will then use the result to deduce what the force of gravity looks like between two solid spheres.

The geometric relationship between the conglomerate of variables described above is depicted in Figure A.2. This will aid us in relating them all. Our first calculation will be to compute the infinitesimal mass of the strip from the spherical shell of thickness ds. Assuming the thickness of the shell ϵ is sufficiently small (so that the radius to the inside and outside of the shell are practically identical at R), the volume V of the entire shell is

$$V = 4\pi R^2 \epsilon.$$

And because the mass of the shell is m_1, the density ρ of the sphere is

$$\rho = \frac{m_1}{V} = \frac{m_1}{4\pi R^2 \epsilon}.$$

Now the slice we take out of the shell is an annulus of width ds, very small thickness ϵ, and radius $R\sin(\theta)$ (see Figure A.2). Therefore, its

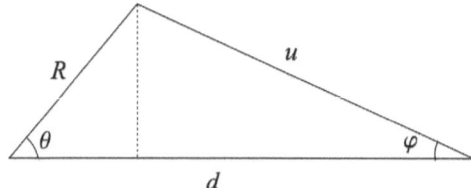

Figure A.2: The geometric relationship between m_1, m_2, and the slice ds with mass dM.

volume element dV is

$$dV = 2\pi R \sin(\theta)\epsilon ds.$$

Using the density ρ derived previously, the mass element dM of the slice is simply

$$dM = \rho dV = \frac{m_1}{2R} \sin(\theta) ds.$$

Notice that the slice ds is a small arc of the sphere of radius R. Hence, if $d\theta$ is a small change in the angle θ, then $ds = R d\theta$ by the arclength formula $s = r\theta$. Consequently, the mass element is

$$dM = \frac{m_1}{2} \sin(\theta) d\theta. \tag{A.5}$$

Let's now analyze what the force on the point mass m_2 by this small mass element dM looks like. Imagine breaking the slice of the sphere into infinitesimal pieces itself, each a distance u away from the mass m_2. Here, you essentially have two point masses tugging on each other gravitationally. But this is precisely the situation to which (A.4) applies. And if we consider the effect of each of these slices of the spherical slice, their net effect will be to pull the mass m_2 towards the center of the spherical mass m_1 due to symmetry. Hence, the overall gravitational effect of the slice is

$$dF = -\frac{GdM m_2}{u^2} \cos(\varphi),$$

§A.3. Derivation of Newton's Law of Gravity

where the $\cos(\varphi)$ accounts for the symmetry of the situation (all the vertical forces cancel, and m_2 is tugged solely in the radial direction towards the center of m_1). Accounting for the mass element dM with (A.5), the force element becomes

$$dF = -\frac{Gm_1m_2}{2u^2}\cos(\varphi)\sin(\theta)d\theta. \tag{A.6}$$

Using Figure A.2, we can express u and thus $\cos(\varphi)$ in terms of d, R, and θ. By the Pythagorean theorem

$$u^2 = [d - R\cos(\theta)]^2 + [R\sin(\theta)]^2$$
$$\implies u = \sqrt{d^2 - 2dR\cos(\theta) + R^2},$$

where we take the positive solution because surly $u > 0$. Because $\cos(\varphi) = \frac{d - R\cos(\theta)}{u}$, we have that

$$\cos(\varphi) = \frac{d - R\cos(\theta)}{\sqrt{d^2 - 2dR\cos(\theta) + R^2}}.$$

Substituting these results into (A.6) modifies the force element to

$$dF = -\frac{Gm_1m_2}{2}\frac{(d - R\cos(\theta))\sin(\theta)}{(d^2 - 2dR\cos(\theta) + R^2)^{3/2}}d\theta.$$

This is quite the algebraic mess. But we will not let it intimidate us. We have successfully deduced what an infinitesimal force element from the slices on the sphere of mass m_2 acting on mass m_1 looks like. To deduce the whole force, we integrate

$$F = \int_0^\pi dF,$$

where the integral is taken from $\theta = 0$ to $\theta = \pi$ because this allows the slices to sweep out the entire sphere (see Figure A.1).

To do this, we must evaluate

$$F = \int_0^\pi -\frac{Gm_1m_2}{2}\frac{(d - R\cos(\theta))\sin(\theta)}{(d^2 - 2dR\cos(\theta) + R^2)^{3/2}}d\theta.$$

We can make this slightly more appealing by factoring out the constants:
$$F = -\frac{Gm_1m_2}{2}\int_0^\pi \frac{(d - R\cos(\theta))\sin(\theta)}{(d^2 - 2dR\cos(\theta) + R^2)^{3/2}}d\theta.$$

To simplify, we make the substitution
$$t = d - R\cos(\theta) \implies dt = R\sin(\theta)d\theta.$$

Notice that with this substitution $R\cos(\theta) = d - t$. Thus, the integral is
$$F = -\frac{Gm_1m_2}{2}\int_{\theta=0}^{\theta=\pi} \frac{t}{(d^2 - 2d(d-t) + R^2)^{3/2}}\frac{dt}{R}$$
$$= -\frac{Gm_1m_2}{2R}\int_{\theta=0}^{\theta=\pi} \frac{t}{(2td - d^2 + R^2)^{3/2}}dt$$

Already this is looking exponentially better. Such an integral is readily solved with integration by parts. (Note, we temporarily neglect to include the constant $-\frac{Gm_1m_2}{2R}$.) Let
$$z = t \implies dz = dt$$
$$dw = \frac{dt}{(2td - d^2 + R^2)^{3/2}} \implies w = -\frac{1}{d}\frac{1}{\sqrt{2td - d^2 + R^2}}.$$

Hence, the integral
$$\int_{\theta=0}^{\theta=\pi} zdw = zw\Big|_{\theta=0}^{\theta=\pi} - \int_{\theta=0}^{\theta=\pi} wdz$$
$$= -\frac{t}{d\sqrt{2td - d^2 + R^2}}\Big|_{\theta=0}^{\theta=\pi} + \int_{\theta=0}^{\theta=\pi} \frac{dt}{d\sqrt{2td - d^2 + R^2}}$$
$$= -\frac{t}{d\sqrt{2td - d^2 + R^2}}\Big|_{\theta=0}^{\theta=\pi} + \frac{1}{d^2}\sqrt{2td - d^2 + R^2}\Big|_{\theta=0}^{\theta=\pi}.$$

Back substituting $t = d - R\cos(\theta)$, we obtain
$$-\frac{d - R\cos(\theta)}{d\sqrt{2(d - R\cos(\theta))d - d^2 + R^2}}\Big|_{\theta=0}^{\theta=\pi}$$
$$+ \frac{1}{d^2}\sqrt{2(d - R\cos(\theta))d - d^2 + R^2}\Big|_{\theta=0}^{\theta=\pi}.$$

§A.3. Derivation of Newton's Law of Gravity

Rather than print all the algebra, we leave it to the reader to verify

$$-\frac{d - R\cos(\theta)}{d\sqrt{2(d - R\cos(\theta))d - d^2 + R^2}}\bigg|_{\theta=0}^{\theta=\pi} = 0.$$

Recalling the forgone factor of $-\frac{Gm_1m_2}{2R}$, the force F is thus

$$\begin{aligned}
F &= -\frac{Gm_1m_2}{2R}\frac{1}{d^2}\sqrt{2(d-R\cos(\theta))d - d^2 + R^2}\bigg|_{\theta=0}^{\theta=\pi} \\
&= -\frac{Gm_1m_2}{2d^2R}\left(\sqrt{(d+R)^2} - \sqrt{(d-R)^2}\right) \\
&= -\frac{Gm_1m_2}{2d^2R}(d+R-(d-R)) \\
&= -\frac{Gm_1m_2}{d^2}.
\end{aligned}$$

This is a momentous result: The gravitational attraction between a spherical shell of mass m_1 and a point mass m_2 is the same—it's as if m_1 were not spherical at all but instead itself a point mass.

Now with knowledge that a spherical mass behaves the same as a point mass, we ought to be able to derive the force between a point mass and a solid sphere. We do this by breaking up the solid sphere into infinitely many shells each of radius r and width dr. If the sphere's total mass is m_1 and radius R, then the density ρ is

$$\rho = \frac{m_1}{V} = \frac{3m_1}{4\pi R^3}.^\dagger$$

The volume element dV of one of these shells, similar to the volume of the shell above with thickness ϵ, is

$$dV = 4\pi r^2 dr \implies dM = \rho dV = \frac{3m_1 r^2}{R^3}dr.$$

Now if the point mass m_2 is a distance d away from the sphere, the gravitational attraction between it and one of the shells (as derived

---

†This assumes the sphere has constant density. Naturally, it need not be (and, if speaking of planets or stars, most likely is not). To analyze these cases, you require a density function $\rho(r)$ over which you would later integrate.

previously) is

$$dF = -\frac{GdMm_2}{d^2} = -\frac{Gm_1m_2}{d^2}\frac{3r^2}{R^3}dr.$$

To deduce the net force, we integrate

$$F = \int_0^R dF$$

as the shells making up the sphere range from radius $r = 0$ to $r = R$. Therefore, the integral we must evaluate is

$$F = \int_0^R -\frac{Gm_1m_2}{d^2}\frac{3r^2}{R^3}dr = -\frac{Gm_1m_2}{d^2}\int_0^R \frac{3r^2}{R^3}dr,$$

which is trivial. The answer is amazingly

$$F = -\frac{Gm_1m_2}{d^2},$$

the same result when m_1 was a point mass. Hence, not only can spherical shells be approximated as point masses in Newton's law of gravity, but solid spheres as well! Hopefully this analysis has revealed to you why Newton's law of gravity is so profound.

§A.4 Digression on the Maclaurin Series for e^x

In the section on Taylor series and Maclaurin expansions we introduced the polynomial representation of the exponential function:

$$e^x = \exp(x) = \sum_{n=0}^{\infty} \frac{x^n}{n!} = 1 + x + \frac{x^2}{2!} + \frac{x^3}{3!} + \cdots. \quad \text{(A.7)}$$

Here we expand upon this formula to derive yet another extraordinary limit for π.

Recall from the section on the Gaussian function that integrating $\exp(-x^2)$ is quite a cumbersome task. Through an abnormal amount of algebra and double integrals, we computed

$$\int_0^{\infty} \exp(-x^2)dx = \frac{\sqrt{\pi}}{2}. \quad \text{(A.8)}$$

§A.5. Reconciling Two Formulas for e

With our fresh understanding of Maclaurin series, we'll evaluate (A.8) with $\exp(-x^2)$ replaced by its Maclaurin series. But what does this series look like? Fortunately, the derivation is quite simple. All we do is substitute $-x^2$ for x in (A.7) so that

$$\exp(-x^2) = \sum_{n=0}^{\infty} \frac{(-x^2)^n}{n!} = \sum_{n=0}^{\infty} \frac{(-1)^n x^{2n}}{n!}.$$

Using (A.8), it follows that

$$\int_0^{\infty} \exp(-x^2) dx = \int_0^{\infty} \sum_{n=0}^{\infty} \frac{(-1)^n x^{2n}}{n!} dx = \frac{\sqrt{\pi}}{2}.$$

Before getting intimidated by integrating over an infinite series, recognize that the notation $\frac{(-1)^n x^{2n}}{n!}$ is intended to characterize all terms that appear in the series. Logically, then, integrating this characteristic term should be emblematic of integrating each term separately, and this is indeed the case. Hence, to integrate the series, we need only integrate the general term $\frac{(-1)^n x^{2n}}{n!}$ and, when expanded, all terms will be formatted as if they were integrated separately. With this, we integrate to obtain

$$\int_0^{\infty} \sum_{n=0}^{\infty} \frac{(-1)^n x^{2n}}{n!} dx = \sum_{n=0}^{\infty} \frac{(-1^n) x^{2n+1}}{(2n+1) n!} \bigg|_0^{\infty} = \frac{\sqrt{\pi}}{2},$$

from which it follows that

$$\lim_{x \to \infty} \sum_{n=0}^{\infty} \frac{(-1)^n x^{2n+1}}{(2n+1) n!} = \lim_{x \to \infty} \left[x - \frac{x^3}{3 \cdot 1!} + \frac{x^5}{5 \cdot 2!} - \frac{x^7}{7 \cdot 3!} + \cdots \right]$$

$$= \frac{\sqrt{\pi}}{2}. \tag{A.9}$$

This result is quite bizarre and is a rather difficult limit to wrap one's mind around. Though many simple limits exist for π, it is somewhat reassuring that there are also more ornate ones, precisely like (A.9).

§A.5 Reconciling Two Formulas for e

Here we would like to derive the remarkable series

$$e = 1 + 1 + \frac{1}{2!} + \frac{1}{3!} + \frac{1}{4!} + \cdots$$

by using the definition

$$e := \lim_{n \to \infty} \left(1 + \frac{1}{n}\right)^n$$

as claimed possible in §4.1. Using what is called a *binomial expansion* (a.k.a. the *binomial theorem*), it can be shown that any two numbers $\delta, \epsilon \in \mathbb{C}$ expressed algebraically with structure $(\delta + \epsilon)^n$ has an expansion detailed by the sum

$$(\delta + \epsilon)^n = \sum_{k=0}^{n} \binom{n}{k} \delta^{n-k} \epsilon^k, \qquad (A.10)$$

where $\binom{n}{k}$ is the binomial coefficient (see §5.3). Setting $\delta = 1$ and $\epsilon = \frac{1}{n}$ in (A.10) yields a series formulation for the algebraic equation we are interested in:

$$\left(1 + \frac{1}{n}\right)^n = \sum_{k=0}^{n} \binom{n}{k} \frac{1}{n^k}.$$

Using the equality

$$\binom{n}{k} = \frac{n!}{k!(n-k)!}$$

we have

$$\left(1 + \frac{1}{n}\right)^n = \sum_{k=0}^{n} \frac{n!}{k!(n-k)!} \frac{1}{n^k}.$$

Eventually we will let $n \to \infty$. But before we do, let's expand this series to get a better handle on things:

$$\sum_{k=0}^{n} \frac{n!}{k!(n-k)!} \frac{1}{n^k} = 1 + \frac{n}{n^1} + \frac{n(n-1)}{2!n^2} + \cdots + \frac{1}{n^n}.$$

Expanding the polynomials in the numerators and taking the limit as $n \to \infty$, the following result is obtained:

$$e = 1 + 1 + \frac{1}{2!} + \frac{1}{3!} + \frac{1}{4!} + \cdots. \qquad (A.11)$$

On top of this being a remarkable formula, (A.11) reassures us that the Maclaurin expansion of $\exp(x)$ in §4.1 is a genuine representation for the exponential function.

§A.6 The Fundamental Theorem of Arithmetic

In this supplement, we prove the fundamental theorem of arithmetic, which we used in §4.5 to prove the infinitude of primes. The theorem states:

Theorem A.2 (Fundamental Theorem of Arithmetic). *All positive integers $n > 1$ can be factored into a unique product of primes (excluding the order in which this prime product is multiplied).*

We prove this in two parts. First we establish that every natural number n can be written as a product of primes. Following this, we prove that this product is a unique one, up to the order in which the primes are multiplied.

Proof: Part 1. We prove the first part with induction.[†] Ignoring one, the smallest composite number is four, which has a prime factorization $4 = 2^2$. No other primes multiplied together yield four, so this is indeed unique. Altogether, the base case is proved. Our inductive hypothesis is to assume that all integers up to and including some positive integer $n > 4$ can be written as a product of primes. We prove the first part of the theorem if we prove $n+1$ is also a product of primes.

It is clear that the number $n+1$ is either composite or not. If it is not, then $n+1$ is prime and we have shown that $n+1 \in \mathbb{P}$ and we are done. If $n+1$ is not prime, then by the definition of a composite number there exists some $d \in \mathbb{N}$ such that $d \neq 1$, $d \neq n+1$, and $d \mid (n+1)$. In other words, $n+1 = kd$ for some $k \in \mathbb{N}$ such that $1 < k < n+1$. But because d and k are less than $n+1$, it

---

[†]See §A.8 for a brief introduction to proofs by induction.

follows from the inductive hypothesis that they can be expressed as a product of primes in the form

$$k = \prod_{i=1}^{r} p_i \text{ and } d = \prod_{j=1}^{s} q_j,$$

where all $p_i, q_j \in \mathbb{P}$. Thus,

$$n + 1 = \left(\prod_{i=1}^{r} p_i\right)\left(\prod_{j=1}^{s} q_j\right),$$

which is a product of primes. □

Proof: Part 2. To prove uniqueness, we assume the natural number β has two distinct prime factorizations:

$$\beta = p_1 \times p_2 \times \cdots \times p_n \text{ and } \beta = q_1 \times q_2 \times \cdots \times q_m, \quad (A.12)$$

where all $p_i, q_j \in \mathbb{P}$. We will prove that these two representations are actually the same.

Because both prime factorizations in (A.12) are those for the same number β, we require

$$p_1 \times p_2 \times \cdots \times p_n = q_1 \times q_2 \times \cdots \times q_m.$$

Naturally, it is possible there exists a few factors in common on both sides. Our main assumption will be that these cancel, so that we have the new equality

$$p_i \times p_{i+1} \times \cdots \times p_{i+\gamma} = q_j \times q_{j+1} \times \cdots \times q_{j+\delta} \quad (A.13)$$

such that all pairs of p and q are coprime (i.e. they share no common divisors—see our discussion concerning Lehmer's totient problem on page 231). We will show this is not the case.

Define

$$Q := q_j \times q_{j+1} \times \cdots \times q_{j+\delta}$$

§A.7. On Grandi's Series

and
$$k := p_{i+1} \times p_{i+2} \times \cdots \times p_{i+\gamma}.$$

It is evident that $Q = kp_i \implies p_i \mid Q$. In other words, we have shown p_i divides Q. This, of course, requires Q to have a factor equal to p_i. Since p_i is prime, Q is therefore required to have a *prime* factor equal to p_i. But this means there is a prime $q_{j+\alpha} = p_i$ in the prime factorization of β, which contradicts our assumption that all the common factors had canceled in (A.13). Consequently, we are forced to conclude that the prime factorizations in (A.12) are actually the same. □

The uniqueness of a number's prime factorization constructs a compelling argument for why the number one should not be considered prime. To illustrate, suppose

$$n = \prod_{k=1}^{k} p_k^{\alpha_k}$$

is the prime-factorization of a number $n \in \mathbb{N}$, where all p_k are distinct primes and $\alpha_k \in \mathbb{N}_0$. Suppose we let $1 \in \mathbb{P}$. Then

$$n = 1 \times \prod_{k=1}^{k} p_k^{\alpha_k} = 1 \times 1 \times \prod_{k=1}^{k} p_k^{\alpha_k} = \cdots = 1 \times \cdots \times 1 \times \prod_{k=1}^{k} p_k^{\alpha_k}.$$

This clearly makes the prime factorization ambiguous, infringing the uniqueness portion of the fundamental theorem of arithmetic. Hence, in order to preserve uniqueness, we simply dismiss one from the set of primes—a small price to pay for such an important theorem.

§A.7 On Grandi's Series

In §4.11, we claim the following series (known as Grandi's series)
$$S_1 = \sum_{n=0}^{\infty}(-1)^n = 1 - 1 + 1 - 1 + 1 - 1 + \cdots$$

converges to $\frac{1}{2}$. Here we provide a more intricate analysis for why this is an appropriate value to assign S_1. Here, our analysis considers the rather intricate integral expression

$$L := \int_0^1 x^{\alpha x^\beta} dx, \qquad (A.14)$$

where $\alpha, \beta \in \mathbb{R}$. (Indeed, we will specify the exact values for α and β later. For now, we seek generality.) Observe that the integrand

$$x^{\alpha x^\beta} = e^{\alpha x^\beta \log(x)} = \sum_{n=0}^\infty \frac{\alpha^n x^{\beta n} \log^n(x)}{n!},$$

which follows from the Maclaurin series expansion for e^x. Hence, (A.14) is equivalent to

$$L = \int_0^1 \sum_{n=0}^\infty \frac{\alpha^n x^{\beta n} \log^n(x)}{n!} dx.$$

Evidently, this looks many orders of magnitude worse than (A.14). But, as we have seen many times, a sneaky substitution can go a long way—this is in our future. In the mean time, notice that by the elementary integral property

$$\int [f(x) + g(x)] dx = \int f(x) dx + \int g(x) dx,$$

where $f(x)$ and $g(x)$ are arbitrary functions, we can bring the integral into the sum so that

$$\sum_{n=0}^\infty \int_0^1 \frac{\alpha^n x^{\beta n} \log^n(x)}{n!} dx = \sum_{n=0}^\infty \frac{\alpha^n}{n!} \int_0^1 x^{\beta n} \log^n(x) dx,$$

where, in the last step, we brought the factor $\frac{\alpha^n}{n!}$ out of the integral because it does not depend on x. Here, each integral is of the form

$$\int_0^1 x^m \log^n(x) dx, \qquad (A.15)$$

§A.7. On Grandi's Series

where $m = \beta n$. We can evaluate (A.15) with a clever u-substitution. Let

$$u = -\log(x) \implies dx = -e^{-u}du,$$

thereby establishing the equality

$$\int_0^1 x^m \log^n(x) dx = (-1)^n \int_0^\infty e^{-(m+1)u} u^n du.$$

You may be getting excited due to the structure of this integral. To make it explicit we incorporate the final substitution

$$\nu = (m+1)u \implies d\nu = (m+1)du.$$

This prompts the equality

$$(-1)^n \int_0^\infty e^{-(m+1)u} u^n du = \frac{(-1)^n}{(m+1)^{n+1}} \int_0^\infty e^{-\nu} \nu^n d\nu.$$

Observe that the integral on the right-hand side is just $\Gamma(n+1)$—the gamma function (see §3.7). Reconciling this with (A.15), we see that

$$\int_0^1 x^m \log^n(x) dx = \frac{(-1)^n}{(m+1)^{n+1}} \Gamma(n+1).$$

And upon transferring this result back into (A.14), we have

$$L = \sum_{n=0}^\infty \frac{a^n}{n!} \frac{(-1)^n \Gamma(n+1)}{(\beta n + 1)^{n+1}},$$

because $m = \beta n$. Here, the variable of summation $n \in \mathbb{N}_0$ making $\Gamma(n+1) = n!$ by our analysis in §3.7.[†] Utilizing this property and the equality in (A.14), we obtain

$$\int_0^1 x^{ax^\beta} dx = \sum_{n=0}^\infty \frac{(-1)^n a^n}{(\beta n + 1)^{n+1}}. \tag{A.16}$$

Notice that with $\alpha = 1$ and $\beta = 0$, (A.14) equates to

$$\int_0^1 x dx = \frac{1}{2}.$$

[†] For $n = 0$ the property $0! = 1$ is implied.

But by (A.16), this implies

$$\sum_{n=0}^{\infty}(-1) = 1 - 1 + 1 - 1 + \cdots = \frac{1}{2},$$

which is Grandi's series.

§A.8 Proofs by Induction

Throughout this text we have encountered many methods of proof, most regularly proofs by contradiction. In these we assume the opposite of what is to be shown and demonstrate our supposition is contradictory, such as $2 = 1$. In this section we walk through another common proof strategy called *induction*.

Before we plunge into the process of inductive proofs, let's first establish something to prove (something, of course, that is well-suited for a proof by induction). Consider the following sums:

$$1 = 1^2$$
$$1 + 3 = 2^2$$
$$1 + 3 + 5 = 3^3$$
$$1 + 3 + 5 + 7 = 4^2$$
$$\vdots$$

In general, it appears the sum of the first n odd integers is n^2. In other words,

$$\sum_{k=1}^{n}(2k - 1) = 1 + 3 + 5 + \cdots + (2n - 1) = n^2.$$

This formula, however, remains at the speculation stage. We noticed a pattern for the sum of the first few odd integers and claimed it to be true in general, though have yet to prove it. As you can guess, such a proof is most conveniently completed using induction. This is

§A.8. PROOFS BY INDUCTION

because inductive proofs are most fitting to formulas whose domain is the natural numbers (or, equally well, whole numbers).

The general outline of an inductive proof is as follows:

- Prove the simplest case (called the *base case*), which involves the smallest value in the domain.

- Assume whatever you are trying to show holds for an arbitrary number n in the domain (this is called the *inductive hypothesis*).

- Prove using the inductive hypothesis and (potentially) the base case that the theorem holds for the case $n+1$.

To understand why this proves a conjecture, consider the most general case of a theorem to which induction is applicable:

Theorem A.3. *If $n \in \mathbb{N}$, then $f(n) = g(n)$.*

Now we have neither specified $f(n)$ nor $g(n)$, and this is because we are trying to be as general as possible. Inductive proofs work well for theorems of this type, as hinted towards in the example above. Rewriting the outline for induction except in the context of this theorem, we have

- The smallest element in the domain is $n = 1$. Hence we must prove $f(1)$ equals $g(1)$. This is a simple algebraic check wherein you compare the values $f(1)$ and $g(1)$. If $f(1) = g(1)$, the base case is proved. Otherwise, you may want to rethink the theorem.

- Assume that for some n, $f(n) = g(n)$.

- Using our knowledge of $f(1)$ and, likely more helpful, that $f(n) = g(n)$, prove $f(n+1) = g(n+1)$.

This proves $f(n) = g(n)$ for all $n \in \mathbb{N}$ because you proved it worked for $n = 1$ and for $n + 1$. Therefore, it must work for $n = 1 + 1 = 2$. But it also works for $n + 1$, so it works for $n = 3$. The argument continues, so it must work for $n = 4, 5, 6, \cdots$. In other words, it works for all $n \in \mathbb{N}$.

To illustrate this more concretely, consider again the theorem posed above:

Theorem A.4. *If $n \in \mathbb{N}$, then*

$$\sum_{k=1}^{n}(2k-1) = n^2. \tag{A.17}$$

Proof. The first step in an inductive proof is to show that the proposed formula works for the smallest value in the domain. For our purposes, this is the value $n = 1$. Substituting $n = 1$ into (A.17), we obtain

$$\sum_{k=1}^{1}(2k-1) = 1 = 1^2 = n^2.$$

Hence, the base case is proved.

We now make our inductive hypothesis—that for some $n \in \mathbb{N}$, (A.17) is true. Exploiting this hypothesis, we will prove (A.17) is true for $n + 1$. Notice the sum

$$\sum_{k=1}^{n+1}(2k-1) = 2(n+1) - 1 + \sum_{k=1}^{n}(2k-1)$$
$$= 2n + 1 + \sum_{k=1}^{n}(2k-1)$$

From the inductive hypothesis we know

$$\sum_{k=1}^{n}(2k-1) = n^2,$$

and so

$$\sum_{k=1}^{n+1}(2k-1) = 2n+1+n^2$$
$$= (n+1)^2.$$

Hence we have shown

$$\sum_{k=1}^{n+1}(2k-1) = (n+1)^2,$$

which is (A.17) with input $n+1$. This is the result to be shown, and so (A.17) is substantiated. □

Hopefully the power of induction has made itself apparent. We guarantee it will serve you well in future mathematical endeavors.

§A.9 The Differential Equation $ay''+by'+cy=0$

In §5.5, we performed an analysis with a goal of computing the time it would take to fall down a hole drilled through the Earth. Upon doing so, we encountered the second-order differential equation

$$\frac{d^2r}{dt^2} = -\frac{GM}{R^3}r,$$

where G is the gravitational constant, $M = M_\oplus$ is the mass of the Earth, $R = R_\oplus$ the radius of the Earth, and r the instantaneous distance from the center of the Earth. As if out of nowhere, we claimed the solution to such a differential equation to be

$$r(t) = A\cos(\omega t) + B\sin(\omega t),$$

where A, B, and ω are constants. Here we prove this assertion by computing $y(t)$ for the more general second-order differential equation

$$a\frac{d^2y}{dt^2} + b\frac{dy}{dt} + cy = 0, \qquad (A.18)$$

where $a, b, c \in \mathbb{R}$ and $a \neq 0$ (if $a = 0$ the differential equation would not be second-order). (A.18) can be expressed more succinctly using prime notation and by dividing out the a:

$$ay'' + by' + cy = 0 \implies y'' + \alpha y' + \beta y = 0, \qquad \text{(A.19)}$$

where $\alpha = \frac{b}{a}$ and $\beta = \frac{c}{a}$ (again, $a \neq 0$ so this division is possible).

Consider, just out of the blue, the quadratic equation

$$\lambda^2 + \alpha \lambda + \beta = 0,^{\dagger} \qquad \text{(A.20)}$$

whose roots are

$$\lambda_1 = \sigma + \vartheta i \text{ and } \lambda_2 = \sigma - \vartheta i,$$

where $i := \sqrt{-1}$ and $\sigma, \vartheta \in \mathbb{R}.^{\ddagger}$

Given the roots of (A.20), we can factor the quadratic expression into a product of linear factors such that

$$\lambda^2 + \alpha \lambda + \beta = (\lambda - \lambda_1)(\lambda - \lambda_2) = 0.$$

Expanding, we have

$$\lambda^2 + \alpha \lambda + \beta = \lambda^2 - (\lambda_1 + \lambda_2)\lambda + \lambda_1 \lambda_2.$$

By comparison, it follows that the constants

$$\alpha = -(\lambda_1 + \lambda_2) \text{ and } \beta = \lambda_1 \lambda_2.$$

Hence, the differential equation in (A.19) expands to

$$y'' + \alpha y' + \beta y = y'' - (\lambda_1 + \lambda_2)y' + (\lambda_1 \lambda_2)y$$
$$= (y'' - \lambda_1 y') - \lambda_2(y' - \lambda_1 y).$$

---

†In the study of differential equations, (A.20) is referred to as the *characteristic* or *auxiliary equation* of (A.19).

‡Here, ϑ is an alternative symbol for θ. The angular meaning of ϑ will be made clear shortly.

§A.9. The Differential Equation $ay'' + by' + cy = 0$

Let
$$u = y' - \lambda_1 y \implies u' = y'' - \lambda_1 y'. \tag{A.21}$$

Then,
$$y'' + \alpha y' + \beta y = u' - \lambda_2 u = 0 \implies u' = \lambda_2 u.$$

This is a familiar separable, first-order differential equation whose solution is found by integrating:

$$\int \frac{u'}{u} dt = \int \lambda_2 dt \implies u = Ce^{\lambda_2 t},$$

where $C \in \mathbb{R}$ is the constant of integration. Back-substituting using (A.21) prompts the differential equation

$$Ce^{\lambda_2 t} = y' - \lambda_1 y. \tag{A.22}$$

Unfortunately, this equation cannot be solved using separation of variables. Instead, (A.22) requires a more involved method based on a special function called an *integrating factor*, denoted $\mu(t)$. The purpose of $\mu(t)$ is to give us a function such that

$$\frac{d}{dt}[\mu(t)y] = \mu(t)Ce^{\lambda_2 t},$$

on which we can perform separation of variables and solve for y.

Recall from the product rule that

$$\frac{d}{dt}[\mu(t)y] = \mu(t)y' + \mu'(t)y.$$

If this is to equal $\mu(t)Ce^{\lambda_2 t}$, then it must also equal $\mu(t)(y' - \lambda_1 y)$ by (A.22). That is, we require

$$\mu(t)y' + \mu'(t)y = \mu(t)y' - \mu(t)\lambda_1 y \implies \mu'(t) = -\mu(t)\lambda_1.$$

But this is just a separable differential equation in $\mu(t)$. Integrating, we find

$$\int \frac{\mu'(t)}{\mu(t)} dt = \int -\lambda_1 dt \implies \mu(t) = De^{-\lambda_1 t},$$

where D is yet another constant of integration. By the way we derived $\mu(t)$, we have established that

$$\frac{d}{dt}\left[De^{-\lambda_1 t}y\right] = \left(De^{-\lambda_1 t}\right)Ce^{\lambda_2 t}$$

$$\implies \frac{d}{dt}\left[e^{-\lambda_1 t}y\right] = Ce^{(-\lambda_1+\lambda_2)t}.^\dagger$$

This, again, is a separable differential equation, except now in y. After integrating, we find its solution to be

$$e^{-\lambda_1 t}y = \left(\frac{C}{-\lambda_1+\lambda_2}\right)e^{(-\lambda_1+\lambda_2)t} + E, \qquad (A.23)$$

where E is the constant of integration. Notice, the value $\frac{C}{-\lambda_1+\lambda_2}$ is just another constant, which we hereafter call F.

Solving for y in (A.23), we secure the solution

$$y = Ee^{\lambda_1 t} + Fe^{\lambda_2 t}, \qquad (A.24)$$

which you are encouraged to verify via the original differential equation in (A.19).

We can expand upon this a little more, because earlier we generalized the roots λ_1 and λ_2 to equal $\sigma + \vartheta i$ and $\sigma - \vartheta i$, respectively. Substituting these into (A.24), we obtain

$$y = Ee^{(\sigma+\vartheta i)t} + Fe^{(\sigma-\vartheta i)t}$$
$$= Ee^{\sigma t}e^{\vartheta i} + Fe^{\sigma t}e^{-\vartheta i}$$
$$= e^{\sigma t}\left(Ee^{\vartheta i} + Fe^{-\vartheta i}\right).$$

By Euler's equation (§3.2 and §4.2) $e^{i\theta} = \cos(\theta) + i\sin(\theta)$ and the trigonometric identities $\cos(-\theta) = \cos(\theta)$ and $\sin(-\theta) = -\sin(\theta)$, we have

$$y = y(t) = e^{\sigma t}\left[(E+F)\cos(\vartheta t) + (E-F)i\sin(\vartheta t)\right], \qquad (A.25)$$

---

†*Note:* We assume $\lambda_1 \neq \lambda_2$. The case where $\lambda_1 = \lambda_2$ requires a separate treatment which the reader is encouraged to pursue.

§A.10. Mersenne Primes and Perfect Numbers

thus revealing the angular meaning of ϑ (it's the argument in the trigonometric functions).

To finish up, we would like to remove the complex unit i from the expression. We can do this by expanding the domain of the constants in (A.25) beyond the real numbers into the complex plane. In other words, we suppose $A, B \in \mathbb{C}$. Then, after noting that both $E + F$ and $(E - F)i$ are constants in $\mathbb{C}$, we can set $A = E + F$ and $B = (E - F)i$ so that our final solution to (A.19) becomes

$$y(t) = e^{\sigma t}\left[A\cos(\vartheta t) + B\sin(\vartheta t)\right].^{\dagger}$$

This equation is ubiquitous in the analysis of oscillatory systems in physics, and is the justification for why gravitationally-induced, oscillatory behavior follows a person's valiant (though imbecile) leap into a hole dug through the Earth.

§A.10 Mersenne Primes and Perfect Numbers

We asserted in §5.7 that there exists a fundamental link between perfect numbers and Mersenne primes. This supplement intends to prove such a connection. As a reminder, a perfect number n is any positive integer for which the sum-of-divisors function $\sigma(n) = 2n$. On the other hand, a Mersenne prime is a prime that can be expressed in the form $2^k - 1$ where k is a positive integer.

The theorem we intend to prove is:

Theorem A.5. *If $\rho = 2^n - 1$ is a Mersenne prime, then the number $\kappa = \rho \times 2^{n-1}$ is an even perfect number.*

Proof. We must show that $\sigma(\kappa) = 2\kappa$, which requires us to first determine the divisors of κ. By definition,

$$\kappa = (2^n - 1)(2^{n-1})$$

---

†Again, you are encouraged to verify that this expression satisfies the original second-order differential equation $ay'' + by' + cy = 0$.

because $\rho = 2^n - 1$. It follows that the divisors of κ are all possible permutations of the divisors of ρ and 2^{n-1}. But because ρ is prime, the only divisors of ρ are 1 and itself. The divisors of 2^{n-1} (a power of two) follow just as easily: They are the powers of two less than or equal to 2^{n-1}, namely: $1, 2, 2^2, \cdots, 2^{n-1}$. The divisors of κ are all permutations of 1, ρ, and the divisors listed above. Arranged into a set $\mathcal{D}$, the divisors of κ are

$$\mathcal{D} = \{\underbrace{1, 2, 2^2, \cdots, 2^{n-1}}_{\rho\text{'s divisor of 1}},$$

$$\underbrace{2^n - 1, 2(2^n - 1), 2^2(2^n - 1), \cdots, 2^{n-1}(2^n - 1)}_{\rho\text{'s divisor of } 2^n - 1}\}.$$

Summing over $\mathcal{D}$, we have

$$\sigma(\kappa) = \sum_{j \in \mathcal{D}} j$$
$$= \left(1 + 2 + 2^2 + \cdots + 2^{n-1}\right)$$
$$\quad + \left(1 + 2 + 2^2 + \cdots + 2^{n-1}\right)(2^n - 1)$$
$$= \left(1 + 2 + 2^2 + \cdots + 2^{n-1}\right)[1 + (2^n - 1)]$$
$$= \left(1 + 2 + 2^2 + \cdots + 2^{n-1}\right) 2^n.$$

The sum $1 + 2 + 2^2 + \cdots + 2^{n-1}$ we recognize as a finite geometric series with a common ratio of two. Therefore,

$$1 + 2 + 2^2 + \cdots + 2^{n-1} = \frac{1 - 2^n}{1 - 2} = 2^n - 1.^\dagger$$

Altogether, we have

$$\sigma(\kappa) = \left(1 + 2 + 2^2 + \cdots + 2^{n-1}\right) 2^n$$
$$= (2^n - 1) 2^n$$
$$= 2(2^n - 1)(2^{n-1})$$
$$= 2\kappa.$$

[†] This formula was derived earlier in §A.2.

§A.10. Mersenne Primes and Perfect Numbers

That is, $\sigma(\kappa) = 2\kappa$. Consequently, κ is a perfect number, which is the result to be shown. $\square$

In summary, we have proved that every Mersenne prime $\rho = 2^n - 1$ has a corresponding even perfect number equal to $\rho \times 2^{n-1}$ (and vise versa). It follows that a proof on the infinitude of even perfect numbers implies the infinitude of Mersenne primes (again, this goes both ways). Hence, proving one instantaneously proves the other. Proverbially, this is two birds with one stone.

Algorithms for Select Conjectures

All algorithms are typeset in Python and should be easily translated to the language of your preference. For sake of clarity, none have been optimized.

Helper Functions

```
def root(n):
    '''Input: Integer n. Output: floor( n^(1/2) ).'''
    return int( n**(1/2) )   # Compute floor( n^(1/2) )

def getDivisors(n):
    divisors = []                             # Initialize
    for num in range(1, root(n) + 1):         # Loop
        if n % num == 0:                      # Divisor?
            divisors += [num]                 # Add to list
            if num != n // num:               # Perfect square?
                divisors += [n // num]        # Add to list
    return divisors

def getPrimesUpTo(n):
    '''Input: Integer n. Output: Primes up to n.'''
    primes = []                               # Initialize
    for num in range(2, n + 1):               # Loop
        divisors = getDivisors(num)           # Get divisors
        if divisors == [1, num]:              # Prime?
            primes += [num]                   # Add to list
    return primes
```

Algorithms for Select Conjectures

Collatz Conjecture

```
def collatz(n):
    '''Input: Integer n. Output: (Hopefully) 1.'''
    if n == 1:                    # If n is one
        return 1
    elif n % 2 == 0:              # If n is even
        return collatz(n/2)
    else:                         # If n is odd
        return collatz(3*n + 1)
```

Twin Prime Conjecture

```
def twinPrimes(n):
    '''Input: Integer n. Output: Twin primes up to n.'''
    twins = []                                  # Initialize
    primes = list( getPrimesUpTo(n) )           # Get primes
    for prime1 in primes:                       # Loop
        for prime2 in primes:                   # Loop
            if prime1 - prime2 == 2:            # Twin prime?
                twins += [(prime1,prime2)]      # Add to list
    return twins
```

Goldbach Conjecture

```
def goldbach(n):
    '''Input: Integer n. Output: A prime tuple.'''
    primes = list( getPrimesUpTo(n) )    # Get primes
    for prime1 in primes:                # Loop
        for prime2 in primes:            # Loop
            if prime1 + prime2 == n:     # Goldbach sum?
                return prime1, prime2
    return 'Input not even'              # If n is odd
```

Perfect Numbers

```
def perfectNumbers(n):
    '''Input: Integer n. Output: Perfects up to n.'''
    perfects = []                 # Initialize
    for num in range(1, n + 1):   # Loop
```

```
        divisors = getDivisors(num)      # Get divisors
        divisorSum = sum(divisors)       # Sum divisors
        if divisorSum == 2*num:          # Perfect?
            perfects += [num]            # Add to list
    return perfects
```

Selected Bibliography

[1] Aigner, Martin; Ziegler, Günter M. *Proofs from THE BOOK.* Springer (fifth edition), 2014. Print.

[2] Bell, E.T. *Men of Mathematics: The Lives and Achievements of the Great Mathematicians from Zeno to Poincaré.* Touchstone (reissue edition), 1986. Print.

[3] Bollman, Mark. *Basic Gambling Mathematics: The Numbers Behind the Neon.* CRC Press (first edition), 2014. Print.

[4] Boros, George; Moll, Victor H. *Irresistible Integrals: Symbolics, Analysis and Experiments in the Evaluation of Integrals.* Cambridge University Press (first edition), 2004. Print.

[5] Buffon, G. "Essai d'arithmétique morale." *Histoire naturelle, générale er particulière,* Supplément 4, 1777, pp. 46-123.

[6] Collier, Peter. *A Most Incomprehensible Thing: Notes Towards a Very Gentle Introduction to the Mathematics of Relativity.* Incomprehensible Books (second edition), 2014. Print.

[7] Conrad, Keith. "The Gaussian Integral." University of Connecticut, 2016, pp. 1-2.

[8] Dudley, Underwood. *Elementary Number Theory.* Dover Publications, Inc. (second edition), 1978. Print.

[9] Euler, Leonhard. "De summis serierum reciprocarum ex potestatibus numerorum naturalium ortarum dissertatio altera." *Miscellanea Berolinensia* Vol. 7, 1743, pp. 172-192.

[10] Euler, Leonhard. "Remarques sur un beau rapport entre les series des puissances tant directes que reciproques." *Memoires de l'academie des sciences de Berlin*, Vol. 17, 1768, pp.83-106.

[11] Guy, Richard K. *Unsolved Problems in Number Theory*. Springer (third edition), 2010. Print.

[12] Hartle, James B. *Gravity: An Introduction to Einstein's General Relativity*. Pearson Education (first edition), 2003. Print.

[13] Kleppner, Daniel; Kolenkow, Robert. *An Introduction to Mechanics*. Cambridge University Press (second edition), 2014. Print.

[14] Nahin, Paul. J. *Inside Interesting Integrals*. Springer (first edition), 2015. Print.

[15] Newton, Isaac. *Newton's Principia: The Mathematical Principles of Natural Philosophy*. New York, Daniel Adee, 1846. Print.

[16] Northshield, Sam. "A One-Line Proof of the Infinitude of Primes." *The American Mathematical Monthly*, Vol. 122, No. 5, 2015, p. 466.

[17] Riemann, Bernhard, "Ueber die Anzahl der Primzahlen unter einer gegebenen Grösse." *Monatsberichte der Berliner Akademie*, 1859.

[18] Riemann, Bernhard. "Ueber die Darstellbarkeit einer Function durch eine trigonometrische Reihe." *Gesammelte Mathematische Werke*, 1876, pp. 213-253.

[19] Scheinerman, Edward R. *Mathematics: A Discrete Introduction*. Brooks Cole (third edition), 2012. Print.

[20] Spivak, Michael. *Calculus*. Cambridge University Press (third edition), 2006. Print.

[21] Temperature data for Null Island (§2.1) obtained from www.timeanddate.com.

[22] Velleman, Daniel J. *How To Prove It*. Cambridge University Press (second edition), 2006. Print.

[23] Viète, François. *Variorum de Rebus Mathematicis Responsorum, liber VIII*, 1593.

[24] Wallis, John. *Arithmetica Infinitorum*. Oxford, 1656.

List of Symbols

$\mathbb{R}$ (real numbers)
$\mathbb{N}$ (natural numbers)
$\mathbb{N}_0$ (whole numbers)
$\mathbb{Z}$ (integers)
$\mathbb{Q}$ (rational numbers)
$\mathbb{P}$ (prime numbers)
$\mathbb{C}$ (complex numbers)
$\in$ (element)
$\notin$ (not element)
$:=$ (definition)
$\implies$ (implication)
$\propto$ (proportional to)
$\pm$ (plus/minus)
$\mp$ (minus/plus)
$\square$ (QED, tombstone)
$\to$ (approaches)
$\lim$ (limit)
∞ (infinity)
$\triangle$ (triangle)
$\overline{AB}$ (line segment)
$\widehat{AB}$ (arc)
dx (differential)
$\frac{d}{dx}$ (derivative)
$\int$ (integral)

$\sum$ (sum)
$\prod$ (product)
$n!$ (factorial)
$n!!$ (double factorial)
$|$ (divides)
$\nmid$ (not divides)
$\oplus$ (Terra, Earth)
$\odot$ (Sol, Sun)
A, α (alpha)
B, β (beta)
Γ, γ (gamma)
Δ, δ (delta)
E, ϵ (epsilon)
Z, ζ (zeta)
H, η (eta)
$\Theta, \theta/\vartheta$ (theta)
I, ι (iota)
K, κ (kappa)
Λ, λ (lambda)
M, μ (mu)
N, ν (nu)
Ξ, ξ (xi)
O, o (omicron)
Π, π (pi)

List of Symbols

P, ρ (rho) $\Phi, \phi/\varphi$ (phi)
$\Sigma, \sigma/\varsigma$ (sigma) X, χ (chi)
T, τ (tau) Ψ, ψ (psi)
Υ, υ (upsilon) Ω, ω (omega)

Index

Acceleration, 69, 71
Analytic Continuation, 178
Angular Frequency, 215
Antiderivative, 57
Antipodal Points, 24
Asymptotic Relationship, 185

Basel Problem, 153
Binet's Formulas, 194
Binet, Jacques, 194
Binomial Coefficient, 198
Binomial Expansion, 244
Binomial Theorem, 244
Birthday Paradox, 140, 144
Black Holes, 223
    Event Horizon, 224
    Schwarzschild Black Hole, 223
    Schwarzschild Radius, 223
    Singularity, 224
    Spaghettification, 226
    Tidal Forces, 225
Buffon's Needle Problem, 110
Buffon, Comte de, 109

Calculus, 17
Calculus of Variations, 102
    Brachistochrone Problem, 81, 103
Cantor's Diagonal Argument, 209
Cantor, Georg, 209
Cardinality, 205
Catenary, 109
Cathetus, 14
Collatz Conjecture, 227
Collatz, Lothar, 227
Combinatorics, 197
Comparison Test, 136
Complex Analysis, 134
Complex Numbers, 10
Composite Numbers, 10, 145
Coprime Numbers, 231
Cycloid, 81

Derivatives, 21
    Product Rule, 54
Differential Equation
    Auxiliary Equation, 254

Integrating Factor, 255
Second-Order, 253
Separable, 65, 71
Differential Geometry, 221
Dirichlet Function, 39
Dirichlet, Peter Gustav Lejeune, 39
Discontinuous Functions, 19
Divergence Theorem, 213
Dummy Variable, 92

Einstein, Albert, 5, 216
Euler's Equation, 66, 131, 132
Euler's Identity, 66, 133
Euler's Number, 31, 150
Euler's Totient Function, 231
Euler, Leonhard, 31
Euler-Mascheroni Constant, 139, 229

Factorial Function, 34, 95
Double Factorial, 102
Fermat's Conjecture, 16
Fermat's Last Theorem, 16
Fermat, Pierre de, 16
Fibonacci Sequence, 192
Flux, 213
Fractal, 44
Functions
Continuous Functions, 19
Differentiable Functions, 21
Fundamental Theorem of Arithmetic, 145, 245

Fundamental Theorem of Calculus
Part One, 62
Part Two, 62

Gabriel's Horn, 77
Gabriel, Archangel, 77
Gamma Function, 96
Gauss' Theorem, 213
Gauss, Karl Friedrich, 91
Gaussian Function, 91
Gaussian Integral, 91
Goldbach Conjecture, 228
Golden Ratio, 191
Gradient, 212
Grandi, Luigi Guido, 180
Gravitational Constant, 70
Greene, Brian, 180
Gregory's Series, 131
Gregory, James, 131

Harmonic Numbers, 135, 229
Harmonic Series, 134
Hyper-4, 189
Hyperbolic Trigonometric Functions, 103

Imaginary Numbers, 10
Inertial Frame of Reference, 68, 218
Integers, 9
Integrals, 56
u-Substitution, 59
Constant of Integration, 58
Definite Integral, 60

Hyperbolic Substitution, 107
Improper Integral, 62
Indefinite Integral, 57
Integrand, 57
Integration by Inspection, 58
Integration by Parts, 96
Integration by Substitution, 59
  Riemann Sum, 60
  Trigonometric Substitution, 93
Intermediate Value Theorem, 23
Interval Notation, 9
Irrational Numbers, 10

Kelly's Criterion, 55
Kelly, John, 55
Kinematic Equations
  Acceleration, 71
  Position, 72
  Velocity, 72
Koch Snowflake, 44
Koch, Helge von, 44

L'Hôpital's Rule, 32
L'Hôpital's, Guillaume de, 32
Lehmer's Totient Problem, 231
Lehmer, Derrick, 231
Leibniz, Gottfried Wilhelm, 68
Limit, 18
Line Element, 221
Linear Factors, 154
Logarithms

Natural Logarithm, 59
Lorentz Factor, 220
Lorentz, Hendrik, 220

Maclaurin, Colin, 127
Mascheroni, Lorenzo, 140
Mathematical Analysis, 122
Mathematical Arguments
  Conjecture, 12
  Corollary, 12
  Lemma, 11
  Proposition, 11
  Theorem, 11
Maxwell, James Clerk, 218
Mercator, Nicholas, 169
Mersenne Prime, 230, 257
Mersenne, Marin, 230
Methods of Proof
  Induction, 194, 250
    Base Case, 195, 251
    Inductive Hypothesis, 195, 251
  Proof by Contradiction, 146
  Reductio ad Absurdum, 145
Minkowski, Hermann, 223
Multivariate Calculus, 212

Natural Numbers, 9
Nested Radical, 174, 187
Newton, Isaac, 5, 68
Null Island, 25
Number Theory, 145, 191

Parametric Equations, 83

Index

Pascal's Identity, 201
Pascal's Triangle, 199
Pascal, Blaise, 198
Perfect Numbers, 229, 230, 257
Photoelectric Effect, 217
Pigeonhole Principle, 142
Power Tower, 189
Prime Counting Function, 185
Prime Number Theorem, 184
Prime Numbers, 9, 145
Probability, 86
Probability Distributions, 87
    Continuous Distribution, 89
    Discrete Distribution, 87
    Normal Distribution, 89
    Standard Normal Distribution, 90
    Uniform Distribution, 89
Products, 125
    Convergent Product, 125
    Divergent Product, 125
    Infinite Product, 125
Python, 260

Quadratic Formula, 63

Ramanujan, Srinivasa, 180, 187
Recurrence Relationships, 46, 161
Relativity
    General Relativity, 217
    Minkowski Space, 223
    Redshift, 225
    Spacetime, 222
    Special Relativity, 217
    Time Dilation, 220
Riemann Hypothesis, 182, 233
Riemann Zeta Function, 175, 184
    Nontrivial Zeros, 233
    Singularity, 178
    Trivial Zeros, 179, 233
Riemann's Rearrangement Theorem, 165
Riemann, Bernhard, 176
Russell's Paradox, 211
Russell, Bertrand, 210

Schwarzschild, Karl, 223
Series
    p-Series, 154
    Mercator Series, 169
    Absolutely Convergent, 165
    Alternating Series, 165
    Basel Series, 153
    Conditionally Convergent, 166
    Convergent Series, 124
    Divergent Series, 124
    Finite Series, 123
    Geometric Series, 124, 235
    Grandi's Series, 180, 247
    Infinite Series, 123
    Integral Test, 154
    Maclaurin Series, 128
    Partial Sum, 135
    Taylor Series, 127
Set Operations, 206
    Intersection, 206

Union, 206
Set Theory, 205
   Naïve Set Theory, 211
   Zermelo Set Theory, 212
Sets, 8, 205
   Bijection, 208
   Countable Sets, 207
   Countably Infinite Sets, 208
   Elements, 8
   Injection, 207
   Null Set, 205
   Surjection, 208
   Uncountably Infinite Sets, 208
Solid of Revolution, 75
Squeeze Theorem, 38, 163
Statistics, 86
String Theory, 180
Sum-of-Divisors Function, 229, 257

Taylor, Brooks, 127
Terra, 70
Tetration, 189
Totient, 231
Turing, Alan, 5
Twin Prime Conjecture, 228
Twin Primes, 228

Unit Circle, 37, 103
Unit Hyperbola, 103

Viète's π Formula, 175
Viète, François, 175

Wallis Product, 159
Wallis, John, 159
Wheeler, John Archibald, 223
Whole Numbers, 9
Wiles, Andrew, 16
Wobbly Table Theorem, 29

Zermelo, Ernst, 212

About the Author

MATTHEW FOX is a sophomore mathematics and physics major at Harvey Mudd College in Claremont, California. He intends to pursue theoretical physics in graduate school, with a focus on either string theory or loop quantum gravity.

www.ingramcontent.com/pod-product-compliance
Lightning Source LLC
Chambersburg PA
CBHW031611210526
45464CB00004B/1521